PAY DIRT

ACKNOWLEDGMENTS

The author is much indebted to the following for having read this book in manuscript form and for having made many valuable suggestions: William A. Albrecht, Chairman, Dept. of Soils, Missouri College of Agriculture; B. F. Lutman, Professor of Plant Pathology, University of Vermont; Richard Pough of the National Audubon Society; Louis Bromfield, author of *Pleasant Valley;* and Sir Albert Howard, who has been an inspiration as well as a guide.

J. I. R.

PAY DIRT

Farming & Gardening with *Composts*

by

J. I. RODALE

Editor of *Organic Gardening* Magazine

Introduction by

Sir ALBERT HOWARD

Author of *The Soil and Health*

ISBN: 978-1-940001-64-7

Originally published in 1945

Sportsman's Vintage Press
2017

INTRODUCTION

A revolution in farming and in gardening is in progress all over the world. If I were asked to sum up in a few words the basis of this movement and the general results that are being obtained, I should reply that a fertile soil is the foundation of healthy crops, healthy live stock, and last but not least healthy human beings. By a fertile soil is meant one to which Nature's law of return has been faithfully applied, so that it contains an adequate amount of freshly prepared humus made in the form of compost from both vegetable and animal wastes.

This revolution in crop production involves making the very most of the earth's green carpet—that marvelous machinery for producing all our food and a great deal of the raw materials needed by our factories. Both units of this natural factory—the green cells of the leaf and the power which drives them (the energy of sunlight)—owe nothing to mankind. They are the gifts of Providence which all the resources of Science cannot copy, still less improve. Mankind can only assist the food factory in two directions. He can look after the soil on which the green carpet rests and in which the roots of crops and the unpaid labour force of the soil—moulds, microbes, earthworms, and so forth—live and work. He can also by selecting crops by plant breeding methods make the most of the energy of sunlight and of the improved soil conditions. But the plant breeder must avoid one obvious blunder. He must not be content with improving the variety only, otherwise his labours will soon lead to the exhaustion of the soil. The improved variety will take more out of the ground and will soon become a boomerang. The plant breeder, therefore, must always be careful not to confine his attention to the variety, but must increase the fertility of the soil at the same time. Such crops will look after themselves, and insect and fungous pests will do little or no damage.

How has the United States of America ministered to the country's green carpet? The answer is provided by the Year Book of the Federal Department of Agriculture of 1938, which was published under the title—*Soils and Men*. In this work the results of a careful appraisal of the cultivated soils of the country were recorded. It disclosed the alarming fact that no less than 253,000,000 acres, or 61 per cent of the total area under crops, had either been completely or partially destroyed, or had lost most of its fertility. This has arisen from misuse of the land which has resulted in widespread soil erosion.

Soil erosion is the natural consequence of the collapse of the compound soil particles, on the maintenance of which the well-being both of the soil population and the crop depends. These compound particles are made up of fragments of mineral matter glued together by specks of organic matter provided by the activities of the invisible life of the soil. These soil organisms have to be constantly fed with fresh supplies of humus, otherwise the soil soon wears out. When we attempt to replace these supplies by means of artificial manures, we accelerate the wearing-out process. Nature in all such cases hits back by leaving the soil an inert mass of mineral fragments, in which the beneficial soil population are deprived of air, water, food, warmth, and shelter. The death of the soil and of its population is the natural consequence. Nature finally removes the ruins by wind or water to form either a desert or new soil somewhere else under the sea.

What has been the effect of this neglect of the soil on the human population? The results are summed up in Alexis Carrel's masterpiece—*Man The Unknown*. In the United States no less than £700,000,000 a year is spent on medical care for dealing with disease of various kinds, much of which would never have occurred had the restitution of the manurial rights of the soil received proper attention.

Failure to look after the soil under the green carpet does not pay. It leads to the destruction of large areas of land; it creates an inefficient population.

All this can be put right if the law of return is followed and all the available vegetable and animal wastes of the country are converted into compost for the soil. Exactly how this should be done, what results on crops, live stock, and mankind will then be observed will be clear from a perusal of this book, the chapters of which I have just read with the most lively interest. Many things impressed me as this book developed. What gave me most pleasure was to discover that Mr. Rodale possesses that priceless quality—audacity—without which progress is never made. With no previous experience of the land and its ways, nevertheless he courageously acquired a farm, learned how to get it into a fertile condition, and then observed the results of compost on his crops, his live stock, and afterwards on himself and on the members of his family. He thus took his own advice before offering it to his countrymen in the pages of this book and of his new journal—*Organic Gardening*—which, as the years and months pass, goes from strength to strength. All this is very refreshing in a world which tends to become more and more superficial, due in large measure to that disease of civilization—fragmentation—by which such intimately related subjects as agriculture, food, nutrition, and health have become split up into innumerable rigid and self-contained little units, each in the hands of some group of specialists. The experts, as their studies become concentrated on smaller and smaller fragments, soon find themselves wasting their lives in learning more and more about less and less. The result is the confusion and chaos now such a feature of the work of experiment stations and teaching centres devoted to agriculture and gardening. Everywhere knowledge increases at the expense of understanding.

The remedy is to look at the whole field covered by crop production, animal husbandry, food, nutrition, and health as one related subject and then to realize the great principle that the birthright of every crop, every animal, and every human being is health.

ALBERT HOWARD.

14 Liskeard Gardens,
Blackheath, London, S. E. 3

1st March, 1945

Contents

PART FOUR ~ *HEALTH OR DISEASE FROM FOOD*

PART FIVE ~ *GOOD AND BAD FARMING PRACTICES*

PART ONE
The Living Soil

I

THE BIOLOGIC LIFE OF THE SOIL

The soil is not, as many suppose, a dead, inert substance. It is very much alive and dynamic. It teems with bacteria, actinomycetes, fungi, molds, yeasts, protozoa, algae and other minute organisms. All are microscopic plants except the protozoa which represent animal life. As a group, these lower plants and animals are referred to as the biologic life of the soil. Although microbes have been studied and worked with by the medical profession and by industry for over seventy-five years they have been largely neglected in agriculture.

This microbial population of the soil is concentrated mainly in the upper four or five inches where the bulk of the organic matter, their food, is to be found. When you get down to three feet you may find only thirty to forty thousand bacteria per gram of soil, whereas in the upper levels they may reach billions. In a very fertile soil the amount of bacteria in an acre may weigh as high as 600 pounds. When they die their decaying bodies turn to humus and enrich the soil. Most of them are beneficent, helpful organisms; a few are not. Normally they live together in a delicate, balanced relationship closely controlled by nature. If soil conditions get out of hand either because of the intrusion of foreign elements (certain strong chemicals, for example) or through lack of proper food, or climatic change, the inter-relationships become strained, and it is then more difficult to grow plants the way nature intended. The Society of American Bacteriologists has classified about 1,000 species of bacteria. Of this number only about 100 are pathogenic, that is, capable of bringing about disease in plant, animal or man.

These microbes are the real manufacturers of plant food in the soil and not only do they make it but in some cases (the mycorrhiza fungi) actually feed it to the plant. Fungi and bacteria perform the very valuable function of breaking down or decomposing organic matter and thus conditioning the soil and its structure. Research workers of the U. S. Department of Agriculture have discovered two ways in which they do this: Decay bacteria secrete a mucus which glues fine particles of soil into masses that resist the washing effect of rains which otherwise would cause soil erosion, and fungi that feed on plants throw out thread-like filaments or *mycelia* which bind together soil particles into larger masses. More commonly known is the fact that nitrogen-fixing bacteria work in the roots of leguminous plants to extract nitrogen from the air. There are undoubtedly many other important soil inter-relationships. Algae may help bacteria. The latter may furnish food for protozoa. It seems to be a well-regulated little society. All it needs is a diet of the kind of food it likes and a few other necessary conditions.

First there has to be sufficient aeration. Most soil bacteria need a certain amount of oxygen. Following the system of agriculture proposed here with its stress on composts, the soil structure becomes comparatively loose and porous, encouraging maximum microbial multiplication. Conversely, systems that depend overly on chemicals, find eventually a hardpacked soil unfriendly to soil organisms.

Moisture is very important but there should not be too much of it. The physical structure of a soil gained by years of organic farming becomes perfect for retention of the proper amount of moisture. With a hard-packed soil you have either a desert-dry condition in which bacteria cannot live, or one too saturated with water, which encourages harmful anaerobic bacteria, that give rise to a condition of putrefaction rather than fermentation in the breaking down of organic material.

The earth's microscopic citizens thrive when it is warm, that is, in a temperature between 70° to 100° F. Here again, a soil constantly enriched with organic matter becomes warmer earlier in the spring than one which has been abused with constant doses of strong chemicals. Experiments at Government stations have shown invariably that the darker a soil is the more warmth it will absorb. It is a generally accepted fact that the constant application of humus to soil will gradually darken it and eventually turn it almost black. As between a very dark and a yellowish soil there may be a temperature difference of as much as 8° to 10°. This matter of heat and light absorption is extremely important because if the soil warms up earlier in the spring, a farmer can get on to his land earlier.

Another condition favorable to bacteria is a neutral or slightly acid soil; fungi can thrive in a more acid one. In forests where the soil is acid you will find fewer bacteria. There the work is performed mainly by fungi.

It is axiomatic that for growing crops the more microbial life in the soil the better. Most soil text-books agree that the more fertile a soil is, the more bacteria and other soil micro-organisms will be found in it. In the same way it can safely be stated that the more organic matter is applied to soil the more soil organisms result.

Sir E. John Russell in *Soil Conditions and Plant Growth* describes a bacterial count at the Rothamsted Agricultural Station in England. In a field treated with farmyard manure 28,860,000 bacteria were counted per gram of soil. Where complete minerals plus ammonium sulphate were used, only 15,100,000 bacteria were present. Lyon and Buckman in *The Nature and Properties of Soils* note that all the ordinary types of algae are greatly stimulated by the application of farm manure. Practically all investigators agree that the application of organic manures stimulates and increases the biologic life of the soil to a much greater extent than chemical or mineral fertilizers.

Let us examine some ordinary agricultural practices and their effect on the soil's minute organisms.

There are certain species of "bandits" in the soil that war on plants. We attack them, but in doing so we cannot wage a selective warfare. For every pathogenic organism we destroy, we kill off ten beneficial microbes. Instead of stimulating the natural activities of this biologic soil society by feeding it with sufficient natural organic fertilizers, we douse it with poisonous sprays and caustic chemicals. There is no dispute about the fact that certain stronger chemicals reduce the extent of the biologic population in the soil. This fact is taken advantage of in medicine when we use antiseptics. Our strong chemical fertilizers, to a certain extent, have the same effect on the soil's microbial elements. In the preservation of certain foods, chemicals, such as benzoate of soda, are used to destroy the bacteria which otherwise would "spoil" the food.[1]

Sterilizing the soil in boxes for producing young seedlings for transplants, such as tomato and pepper plants, is a questionable practice. It may kill off the organism responsible for "damping off" the young plant, but it also kills many other necessary members of the microscopic life of the soil, which may result in weakened seedlings.

Many farmers who rely chiefly on chemical fertilizers grow green manure crops which they plow under for organic matter. Such crops may consist of rye-grass, clovers, buckwheat, etc. But this is not always as effective as it sounds. The "digestive" powers of a soil—its ability to break down organic matter—depend upon its fertility. Presence of large amounts of raw plant matter can inhibit the actions of bacteria, fungi and earthworms, and thus reduce the digestive power of the soil. Where a young crop of rye-grass is plowed under there may not be much difficulty, but if it gets too high before this is done, the new crop sowed may suffer from an insufficiency of available nitrogen and other plant nutrients.

Plowed under green matter must be worked on by soil bacteria and fungi and to do this they have to feed on the available nitrogen. In an infertile soil it does not take much to use up all available nitrogen, thus the new crop that has been sown suffers. During the next year, when the soil micro-organisms die and their nitrogen-enriched bodies go back to the store of fertility reserves, good crops ensue. Of course,

[1] For the purposes of this book, the terms "chemical fertilizer" or "chemical" shall be taken to mean those commercial or synthetic substances extensively used in the last fifty years in crop production, which in combination with elements in the soil tend to produce insoluble salt residues that are detrimental to fertility. For instance, superphosphate, ammonium sulphate, calcium cyanamid, nitrate of soda, and such combinations as those popularly labeled 8-10-6, etc.; also poison sprays such as the arsenicals. In contrast, ground limestone, dolomite and other forms of natural lime, and the ground phosphate rocks, which are chemicals, strictly speaking, are not to be considered "chemical fertilizers" when we use the term.

a thin layer of green matter will be more or less broken down by an average soil, but heavier crop residues while they will ultimately enrich a soil, will only act as a deterrent to the growth of the next crop. This is one very good reason for applying composted, or predigested organic matter.

In the wheat country of eastern Washington the only humus that is usually returned to the soil is the straw residue of the crop. It is common knowledge there that this practice reduces the succeeding crop yield. Where soil has been thoroughly enriched with organic composts, year after year, the fertility increases to such an extent that its powers of digestion increase enormously. This type of soil has so large a store of nitrogen, that in breaking down new organic matter, the microbes use up only a small portion of it. More than enough is left for the next crop.

A farmer in Kenya, Africa, wrote Sir Albert Howard that he had got his plantation so fertile "the soil would eat almost anything thrown at it from a gunny bag to corn stover." This thought is well described by Eve Balfour in *The Living Soil:*

> "In an experiment designed to measure this factor cotton-wool pads of known weight were buried for four months in untreated Wareham soil (a poor, worn-out soil), in ordinary woodland soil, and in Wareham soil plus C 5 compost. At the end of the period what was left of these pads was dried and reweighed. Representative results based on many repetitions of the experiment showed that in untreated Wareham soil only 10 per cent of this cellulose had been decomposed, in the woodland soil the figure was 33.6 per cent, but in Wareham soil plus compost the percentage of decomposition was over 91 per cent."

The power of "digestibility" of a soil is in direct relation to the amount of microbes, and other life contained in that soil. Chemicals that destroy the beneficial bacterial life lower this power of "digestibility" and make the soil less fertile.

2
THE EARTHWORM

The common earthworm is of vital importance in maintaining soil fertility. It aerates and pulverizes the soil. It actually *produces* topsoil. Without its aid soils would be relatively hard-packed. The earthworm is nature's plow. It bores into the soil and keeps it well aerated, a condition which encourages soil microbes to multiply. Such tunnelled earth permits water to penetrate rather than run off and to maintain the moist condition so necessary to plant life.

About 1881 Charles Darwin, the naturalist, brought out a book entitled *Vegetable Mold and Earthworms.* It was an exposition of the results of years of investigation as to the part earthworms play in nature's scheme of things and concluded that without the earthworm, vegetation would degenerate close to the vanishing point. Unfortunately, the public came to think of one subject only—evolution, when the name Darwin was mentioned and his earthworm book has remained practically unopened on library shelves for over fifty years.

According to Darwin, worms in making their burrows, swallow an enormous quantity of earth out of which they extract any digestible matter it may contain. They also consume fresh and half-decayed leaves, and other organic matter. The leaves are first dragged into the mouths of their burrows to a depth of from one to three inches and are then moistened with a secreted fluid which hastens digestion.

Darwin estimated that more than ten tons of dry earth per acre annually passed through their digestive systems, so that practically all top-soil was "treated" by them every few years. Wonderful soil cultivators, they swallow not only finished soil but also tiny rock particles which they grind up and further reduce with their digestive juices. In thus accelerating the process by which rocks break down into soil compounds, they help to create soil fit for the most exacting gardener.

These little animals work mostly in the top layer of soil although they are occasionally found seven and eight feet below the surface.

They bring to the top basic mineral substances which when broken down release needed elements and add to the fertility of the top-soil. This is especially valuable where soils have been mined through single cropping. They aerate the soil and allow oxygen to penetrate to the roots of plants. Without this oxygen in the soil, growth processes of plants could not possibly take place. With reduced amounts of oxygen, as in the case of hard-packed soils, many of our most important plants are apt to become sick and consequently easy prey to disease and insects.

Earthworms are also extremely valuable to the farmer because they destroy the larvae of certain noxious insects. The California Earthworm Farms conducted an interesting experiment showing that earthworms are an important factor in plant health. They placed nematode-infected plants in cans half of which contained earthworms. In about a year all of the cans in which earthworms had been placed showed practically a 100% clean-up of the nematodes. In the other cans the condition was the same as at the beginning in some cases and worse in others.

Dean William A. Schoenfeld, Director of Agriculture at Oregon State College, in a paper presented to the Oregon State Horticultural Society, December 12, 1942, declares:

> "I have personally observed during my travels in England and the Continent the rejuvenating effect of earthworms on permanent pastures. Some of these pastures were hundreds of years old. They were heavily grazed and only lime or marl and manure, both solid and liquid, were added as fertilizer. I was told that the carrying capacities of the pastures observed were much greater than they were 50 or 100 years ago. Nearly every farmer visited gave much of the credit for the heavier yields to the earthworm."

British experts have estimated that in the intensive six months cotton growing season following the over-flowing of the Nile, earthworm castings mounted to almost 120 tons per acre. This would seem to indicate an earthworm population of about 1,500,000 to the acre. Such a tremendous quantity of earthworms is made possible by the organic material (their food) which the overflowing Nile deposits on the land. The Arabs who live along the banks of the Nile are an extremely healthy race, strong of body and handsome of physique. Their food, raised in soil enriched by these castings, is no doubt one of the primary factors.

Darwin estimated that on a fertile soil an average of 1/5 of an inch in thickness of topsoil per year is added to the surface by means of these castings.

Sir E. J. Russell, late director of the famous Rothamsted Agricultural Experiment Station in England found a direct relationship between the amount of earthworms present in a soil and the application of farmyard manure. In his book, *Soil Conditions and Plant Growth* he states that in soil where no manure was applied only about 13,000 earthworms per acre were found, whereas where substantial amounts of dung were plowed under, over 1,000,000 were counted. In the same book, Russell goes on to state:

> "Where earthworms are active in the soil, organic matter is distributed throughout the layer in which they operate, but where, in cool climates earthworms are few or absent, there is much less mixing; the dead vegetable matter accumulates on the surface, becoming a partly decomposed, acid, peaty mass, in which the normal soil decompositions are not completed."

Where strong chemical fertilizers are used, conditions distasteful to earthworms arise and their numbers rapidly decrease—even to the vanishing point. Ammonium sulphate, a fertilizer that is extensively used by farmers, is particularly harmful to these soilworkers. The U. S. Government itself publicizes this fact by recommending ammonium sulphate as a specific where earthworms are to be killed off, as in the case of putting greens on golf courses. The following, from *Farmer's Bulletin No. 1569,* gives some light on this point:

> "The results of three years' application of ammonium sulphate to sod on the experimental farm of the Department of Agriculture at Arlington, Va., for fertilizing purposes, have shown incidentally that earthworms were eliminated from the plots where this chemical was used. When applied to soils which are naturally neutral or slightly acid in character, this fertilizer creates a strongly acid condition that is distasteful to the worms and they disappear."

Many other chemical fertilizers are slowly but definitely killing off the earthworm population. This was demonstrated at the Research Laboratory at Dornach, Switzerland. Experiments there showed that earthworms did not like soil saturated with artificial fertilizer and if given an opportunity chose earth fertilized with compost, in preference even to soil that was not fertilized at all. If you are interested in seeing how even a mild acid such as vinegar will affect an earthworm, douse one with some. It will cause instant death.

Strong insect sprays containing lead, arsenic, or copper, lime-sulphurs and tar oil, etc., destroy earthworms. In tracts of potato-growing land where these sprays are periodically applied to the land, earthworms are hard to find. What is equally as bad is the fact that much of the bacteria population is adversely affected too, with the frequent result that the soil becomes almost sterile and the farmer has to work in a dead medium. In such cases each succeeding year requires the use of progressively more spray and more chemical fertilizer to get the necessary yield. Similarly, in vineyards or orchards which have been intensively treated with sprays for many years, earthworms are very scarce. In such places the earth becomes hard-packed and extremely difficult of cultivation.

Where any one item in nature's cycle is disturbed it will be found that others are automatically affected. Nature consists of a chain of interrelated and interlocked life cycles. Remove any one factor and you will find that she cannot do her work efficiently. Remove the earthworm, and bacteria fail to thrive.

The question is often asked, whether earthworms eat the roots of plants, especially the fine root hairs. So long as there is humus or organic matter present there will be no eating of roots. Where the farmer or gardener permits the soil to become so sterile of organic matter that the earthworm in desperation turns to the roots for sustenance, there will arise also other more evil consequences. On the other hand, where there is sufficient humus in the soil, root growth will become very vigorous and the number of fine root hairs will far outnumber those on plants growing in a soil that contains less organic matter.

Earthworms can work in almost any kind of soil, even clay. Too much alkalinity has a harmful effect on them. Therefore, one must be careful not to overlime his land. In the winter when the ground is frozen, the earthworm's activity is considerably decreased, but it undoubtedly continues below the frost line.

A soil thoroughly worked by earthworms will absorb a two-inch rainfall in fifteen seconds due to its porous consistency and sponge-like structure whereas in its neighboring clay soil unworked by earthworms, as long as two hours may be required for the same amount to sink in. Every earthworm burrow-hole is a watering tube. The presence of earthworm castings in a pasture is a sure sign that the soil is sweet. Many roots use the worm's long tunnels as a means of getting to lower levels. As these tunnels are lined by the worms with a fertile, liquid form of cast, the roots benefit accordingly.

As the life of the earthworm lasts but a bare year or two, their dying and decaying bodies furnish a substantial amount of excellent fertilizer. In many cases this is no little quantity. The weight of earthworms on a fertile farm which contains about one million earthworms per acre would be about 1,100 pounds per acre.

Earthworm castings are the finest form of humus known. These castings consist of soil and other matter which passes through and mixes with a secretion in their intestinal canals and is of extremely high fertilizer value. At the Connecticut Experimental Station it was found that the nitrogen in these castings is almost five times greater than in ordinary topsoil, the phosphate seven times greater, potash eleven times and magnesium three times. California florists pay a high price for earthworm castings. They report this material as the best they have been able to secure for raising flowers.

Some of the value of earthworms to growing plants has been known since the dawn of history. Sir Bernard Greenwell is authority for the statement that primitive nomad tribes in Central Africa always pitched their camps on ground covered with earthworm casts, as they found this was the best grazing land. Similar things can be found from the folk-lore of farming in many countries.

Dr. Thomas J. Barrett in his book, *The Earthmaster System,* draws attention to the acceptance in conservative agricultural circles of the fact that earthworms have brought about tremendous increases in yields of crops. He quotes from Dr. E. W. Hilgard's *Soils* the interesting Wolney experiment:

"Wolney has shown by direct experimental culture in boxes, with and without earthworms, surprising differences between the cultural results obtained, and this has been fully confirmed by the subsequent researches of Djemil. In Wolney's experiments, the ratio of higher production in the presence of worms varied all the way from 2.6 percent in the case of oats, 63.9 percent in that of rye, 135.9 percent in that of potatoes, 140 percent in vetch, and 300 percent in that of the field pea, to 733 percent in the case of rape."

Dr. Barrett also quotes a letter from a practical earthworm culturist, a Georgia farmer, Mr. R. A. Caldwell, who reports:

"I have planted moss rose in experimental pots, same age and condition, one pot with worms, one without; invariably, the one with the worms will take on new zest and life, and I have had them make such wonderful growth as 16 to 1. Petunias in boxes, I have also grown in such size and profusion as to be unbelievable to one who never had a demonstration of the earthworm's fertilizing and cultivating ability. Petunias in soil of identical fertility with the aid of hundreds of worms burrowing about their roots, produced leaves 1½ to 1¾ inches wide by 3 inches long, while those in the boxes without the worms were ½ inch wide by 1 to 1¼ inch long; and the worm-fertilized plants were many times as tall as the others."

Practically all agricultural text-books on the subject of soil have a kind word to say about the earthworm, but rarely if ever do they dwell on the pernicious effect of chemical fertilizers on these helpful creatures. There is no question that caustic chemicals are responsible for the death of hundreds of billions of earthworms. But when the subject of earthworms is brought up in conversation with the average agronomic scientist, while he will as a rule mildly agree that this little animal is of help to the soil, his passive attitude will clearly indicate that he thinks it more properly a matter for study in a biology class.

Breeding Earthworms

The late Dr. George Sheffield Oliver of Texas, descended from the inventor of the steel plow, James Oliver, is responsible for the popularization of earthworm breeding. About thirty years ago he came across Darwin's book on the earthworm and became interested. It did not take him long to discover that Darwin had strongly underestimated the importance of the earthworm to agriculture. In one of his experiments, he worked with large flower-pots; some of them he painted red, and others green. He removed all worms and worm eggs from the soil in the red ones and placed earthworms in the green pots. The plants in the latter were so patently superior to those in the red pots that Dr. Oliver decided to find out how to make earthworms breed under controlled conditions. When he impregnated his gardens and fruit orchards with these little earthworkers the results were so spectacular that neighbors thought he had some magic formula. His fame spread. Soon he gave up his medical practice and took to landscaping. With the aid of large quantities of earthworms which he bred, he developed beautiful parks and gardens in the most difficult kinds of soil. Some of his larger contracts came from movie people in Hollywood and as his reputation skyrocketed, he finally moved to California where he developed the California Earthworm Farms and put earthworm breeding on a commercial basis.

If you desire to raise earthworms, it is best to buy a box or two of breeders from one of the regular farms. Very poor results will be obtained if you dig your own and try to breed them in boxes.

Raising earthworms serves an important purpose to the vegetable gardener who has heard about the value of compost too late in the fall to make it for the following spring, or to someone who has used up all his compost during the fall. Both can breed earthworms during the winter in boxes of mixed green matter and manure. The worms will break these down in the boxes so that both the compost and the earthworms can be deposited in the rows where planting is to be done in the spring. Not only is this the finest kind of humus, but the earthworms will be a wonderful addition to the garden. An excellent way of securing large quantities of earthworms is to make a thick mulch of hay, straw, corn-stalks, and other materials, and apply it to as large an area of land as you care to experiment with. We covered an area about thirty by twenty feet and made the mulch about fifteen inches high. The effect of the mulch is to maintain a moist condition in the soil and provide the optimum feeding conditions under which earthworms multiply. Within a few months you can, upon digging into the soil, begin to see the effects. In about a year's time there will be an enormous increase in their numbers.

Vegetable gardeners who want to fallow half their ground each year might cover it with such a mulch, the under part of which will decay and enrich the soil in addition to encouraging the worms. Alternate areas each year.

Earthworms may be fed to poultry, placed around fruit and ornamental trees, in vegetable and flower gardens or sold to fishermen. The question arises whether it is practical to breed earthworms on a large scale for use on farms. To the writer's mind, for general farming purposes, it is more advisable to concentrate one's energies on the production of as much compost as possible. The yearly application of this humus to the soil will encourage earthworms to multiply in great and sufficient numbers. To breed them in such a case would be like taking coals to Newcastle.

The best known large-scale earthworm breeders are: Thomas J. Barrett, Roscoe, California, California Earthworm Farms, Ontario, Calif., and Ohio Earthworm Farms, Worthington, Ohio. The School of Horticulture, Ambler, Pennsylvania, can supply earthworm castings by the quart for house plants and window boxes.

3
THE MYCORRHIZA FUNGI

We have discussed the work of soil bacteria and earthworms in creating and mantaining soil fertility. A third great factor in the growth of plants is the contribution made by the tiny fungi growing in the soil.

Biologists in the past have noted that the roots of many plants were infected with microscopic fungi. Commonly these were considered harmful—parasitic or competitive. But modern soil scientists and biologists—principally English—who have done painstaking research and experimental work in this field have discovered that these fungi serve the host plant in a remarkable way, and are in fact indispensable to its well-being.

Dr. William F. Ganung, in his *Textbook of Botany,* describes the function of these mycorrhiza (from *myces,* fungus, and *rhiza,* root) as follows:

"*Fungi,* of certain small kinds, develop in contact with the tips of roots of many plants, particularly those living in much humus, weaving around them a close cover of mycelial threads, which replace the root hairs. This mycorrhiza, as it is named, absorbs water and mineral matters which it transmits to the roots; and there is some reason to believe that it also absorbs soluble organic matters set free in decay of humus but useful again to the plants. The association seems clearly beneficial both to fungus and flowering plants; and accordingly we have here one of the cases where two different organisms derive benefit from their association, a condition called *symbiosis.*"

The mycorrhiza is not a parasite—it does not gain its sustenance from the roots of the plants which it covers but rather it is in partnership with them. Together—the roots and this covering of mycorrhizas—they are able to draw in sustenance from the surrounding soil.

There are two large groups of mycorrhizas, those which encase the roots, and those which invade the root cells of plants. Both are beneficial and of unusual importance to agriculture.

Professor Wilhelm Pfeffer in 1877 was the first to notice the symbiotic relationship between the roots and these fungi. Another German botanist who spent a great deal of time studying the phenomenon of the mycorrhizas was Dr. B. Frank. There were a few others but the whole subject remained in the sphere of scholarly botanical research until comparatively recently. What the early investigators did not know, or if they knew, did not reflect in their writings, was that eventually in the growing process the fungous threads of the mycorrhiza are digested or consumed by the plant through the roots. The digested product enters the sap-stream and helps to mature the plant, for the fungus is very rich in both proteins and carbohydrates.

Dr. M. C. Rayner and her assistant, Dr. Ida Levisohn, have been experimenting with the mycorrhiza for many years at Bedford College in London and Wareham Forest, Dorset, and they have worked in close association with Sir Albert Howard because of the way his investigations tie in with the actions of this group of fungi. Dr. Rayner's work on the mycorrhiza association in tree growth is becoming recognized as of epochal importance to forestry.

Howard, in his book, *An Agricultural Testament*, finds:

"The mycorrhizal association to be the living bridge by which a fertile soil (one rich in humus) and the crop are directly connected and by which food materials ready for immediate use can be transferred from soil to plant. How this association influences the work of the green leaf is one of the most interesting problems science has now to investigate. Is the effective synthesis of carbohydrates and proteins in the green leaf dependent on the digestion products of these soil fungi? It is more than probable that this must prove to be the case. Are these digestion products at the root of disease resistance and quality? It would appear so. If this is the case it would follow that on the efficiency of this mycorrhizal association the health and well-being of mankind must depend."

Probably everyone who reads this book has had difficulty in transplanting rhododendrons, hemlock and other forest evergreens, and has wondered why the contrary things would not grow in soil that was carefully made acid for them, and shade provided. The likeliest answer is that, transplanted, the mycorrhizal association was disturbed, and in the new soil could not be established. If you dig in the forest earth at the base of such trees and shrubs, growing naturally, you will note that the soil for a couple of inches, or even much deeper, is a half-rotted leaf-mold, fibrous, light as peat-moss and of about the same texture, and filled with a grayish mould. And in this layer are many of the surface roots of the plant you are digging up—but you discard this surface "earth." It doesn't seem substantial. If, by chance, you mulch your transplanted shrub with it, you will probably have no difficulty in making it live, for you will have taken its friendly mycorrhiza fungi along with it, and the valuable partnership can be continued in the new soil, with proper mulching, and soil maintenance.

The fact that compost fertilization stimulates the growth of mycorrhizas in the soil is one of the strongest reasons for urging its use, for "crops grown with compost, or ample quantities of farmyard manure," as Lady Balfour points out, in analysing Dr. Rayner's work and that of Sir Albert Howard, "always showed maximum mycorrhizal development, in marked contrast to those grown with artificials."

It was in 1937 that Sir Albert Howard, while Director of the Institute of Plant Industry at Indore, discovered the tremendous value of this plant-fungus association. He found that where plants were grown with artificial fertilizers the "mycorrhizal relationship was either absent or poorly developed." In tea plants, great mycorrhiza formers, even where there was plenty of humus in the soil, he noticed parasitic fungus growths where artificial fertilizers had been used. He tells how steps were taken during 1938 and 1939 to have the roots of sugarcane examined in order to test this point of view:

"Material was obtained from India, Louisiana, and Natal. In all cases the roots exhibited the mycorrhizal association. The large amount of material sent from Natal included canes grown with artificials only, with humus only, and with both. The results were illuminating. Humus is followed by the establishment of abundant mycorrhiza and the rapid digestion of the fungus by the roots of the cane. Artificials tend either to eliminate the association altogether or to prevent the digestion of the fungus by the roots of the cane. These results suggest that the change-over from manure to artificials is at the root of the diseases of the cane and is the cause of the running out of the variety."

Nearly all cultivated plants have the mycorrhizal partnership. It has been noted in wheat, potatoes, rye grass, alfalfa, practically all fruit trees, rubber, coffee, tea, legumes, sugar-cane, banana, strawberries, tobacco, pasture grasses and many others. Orchids cannot grow without the help of mycorrhiza except under certain artificial conditions when high concentrations of sugar or chemicals are fed in the early stages.

Every indication points to the fact that only where the soil is rich in humus will you find well-developed mycorrhizas. Dr. Paul Sorauer in *A Popular Treatise on the Physiology of Plants,* emphasizes that: "the symbiotic association of these green plants with a fungus is only formed when they grow in humus, and is more largely developed as the soil becomes richer in humus."

Some years ago while traveling through the grape-growing regions of France, Howard was on the lookout for healthy growing vines similar to those found in central Asia. After a long search he found some near the town of Jouques in Bouches dû Rhone. Upon questioning the proprietress he discovered that no artificial fertilizers had ever been used there and that they had an excellent reputation for the quality of their wines. He had some of the roots examined and found that they had the mycorrhizal association. Like the Asiatic vines, also cultivated with farmyard manure, they were free from disease.

Professor A. L. McComb of the Iowa State Experimental Station found that he was unable to make transplanted pine seedlings "take" in the new location unless their roots contained mycorrhizas (Research Bulletin No. 314, April, 1943). He also showed that the seedlings that had the mycorrhizas were four times as rich in phosphorus as those which did not have it. He demonstrated that the action of the mycorrhizas unlocked phosphorus in the soil that otherwise would not be available.

In Western Australia, nursery workers found that to have good results in planting tree seedlings in a new nursery they had to give the new soil a dressing of earth from an old nursery. In this way they inoculated them with mycorrhiza.

It is fitting to include here an excerpt from an article by L. F. Easterbrook in the *News Chronicle* of April 15, 1944 on this subject.

"Wareham Heath is as uninviting a tract of country for growing anything as can be found. It is the 'Egdon Heath' of the Hardy novels and Thomas Hardy used its austere unproductiveness to express an unchangeability that even the hand of man was powerless to alter. Today it is being altered by the hand of man, or rather through the researches of a woman scientist, Dr. M. C. Rayner.

"The difficulties there are many. Exposure is severe, large parts of the Heath are difficult to drain, and the soil is not only extremely infertile but is actually poisonous in its untreated state to the growth of beneficial soil fungi.

"Dr. Rayner's problem was how to improve soil conditions so as to permit the development of these fungi. She experimented with various composts, the best of which so far is one made with brewery hop waste and dried blood; another satisfactory one is straw and dried blood. Other materials being experimented with are bracken, heather and sawdust. Artificial manures, such as sulphate of ammonia, are no substitute for the compost, for they tend to prevent fungal growth. Some of these fertilizers, superphosphate for example, have proved fatal to the trees. By using suitable compost, however, trees are growing on Wareham Heath with complete success.

"In the experimental plot I saw both Scots and Corsican pines, the seed of which was sown with compost, that were robust, vigorous in growth and about fifteen feet high. Beside them were spindly weaklings, sown without compost on the same day, whose height could be measured in inches. Britain is already leading the world with this new knowledge gained at Wareham Heath, but now further progress is being made.

"It is an expensive business to compost a whole forest, so they are trying the effect of composting seed beds in a nursery on the Heath and then planting the trees out without further treatment. So far this is succeeding.

"Plants grown on compost-treated soil and then put out on the sterile soil of the Heath are flourishing, while those planted out beside them, grown in an ordinary nursery without compost, are a pretty sorry sight. In fact, better results are being obtained on the infertile heath soil by using Dr. Rayner's methods than on normal agricultural soil. It would appear that if the seedling 'learns' in the nursery how to forage for itself by developing these rootlets for entering into fungal association with the soil, it retains this habit through life and converts food to its use even in soil conditions as bad as those at Wareham. This means an immense saving of expense and labor, for it would only be necessary to compost a few acres of forest nursery instead of thousands of acres of forest.

"Dr. Rayner emphasises that her work only concerns trees. But shall we find that this principle of root and fungus reaction extends far more widely through the vegetable kingdom? Will it enable us to grow better and healthier farm and garden crops? Is it yet another reverse for the school who believe that doping the soil with chemicals is the way to compel nature's co-operation in food production?"

Sir Albert Howard's extensive work with compost-farming on a large scale, observing its use with such widely different crops as tea, coffee, bananas, grapes, cotton, hops, strawberries and the cereals and legumes, indicates the value of mycorrhizas to the farmer, and the necessity for agricultural experiment stations to study this relationship in connection with all our field crops, and to investigate the whole field of plant ecology and soil biology. The work of Rayner, Howard, Balfour and others is brilliant pioneering, scientifically based, but much work probably still remains to be done.

4

THE PENICILLIUM MOLD

One of the soil micro-fungi, of whose past activities and life we are densely ignorant, but which is an associate of the mycorrhiza-forming species is the penicillium mold which has recently made headlines in medicine. About fifteen years ago, an observing English scientist, Dr. Alexander Fleming, by accident discovered that when this mold came in contact with a culture of bacteria a clear space formed around it. His curiosity aroused, he performed experiments which led to the discovery that a substance, penicillin, secreted by this soil organism possessed amazing powers.

Penicillin is such a powerful antiseptic that when one part of it is diluted in two million parts of water it can be used to prevent the multiplication of bacteria. It does not kill them directly. It merely destroys their effectiveness by various tricks. It surrounds them with so much oxygen that they "burn" up. In other cases (that is, other types of the same species of mold) it withholds the oxygen and the bacteria "suffocate." The white corpuscles of the blood then have an easy job of disposing of them.

Penicillin is even more spectacular in its cures than the sulfa drugs and evidently much safer. It does not produce the toxic effects of the sulfas, which in many cases leave their mark on certain organs of the body. Every day new honors are heaped on this mold, as new discoveries widen its uses. It is being tried out as a specific in treating cancer and the laboratory results offer encouragement.

It would seem logical that the highest quality penicillin should be devoted to saving lives. If the corn now used largely in its manufacture were grown in soil enriched exclusively by organic fertilizers, to the total exclusion of strong artificials, would the penicillin secreted by the parent penicillium mold be of better quality than otherwise? Would production be greater also?

Another unusual discovery in the penicillin field is discussed in *Nature* (London). Two physicians, Enoch and Wallersteiner, describe a method of putting the mold itself into the body to permit it to work there, rather than injecting the penicillin which it produces. Many advantages were found from such procedure. This is a much more encouraging trend than the recent attempt to prepare synthetically in chemical form the artificial "equivalent" of the substance that the penicillium mold secretes and thus do away with the need for the mold. This is like trying to produce white corpuscles for sale in drugstores. Penicillin is a biologic, organic substance. No chemist has ever succeeded in producing life and it is probable that none ever will.

The sensational results obtained with penicillin have encouraged many medical investigators to consider other soil micro-organisms, and not without results. One of the first such discoveries was gramicidin, a drug which is extracted from a variety of soil bacteria. Its potency is said to be between 1000 and 100,000 greater than sulfanilamide. Another is H-1 which also is extracted from soil bacteria and has been found effective on war wounds. Parke, Davis and Co., in introducing tyrothricin, had this to say in their *Therapeutic Notes* (Feb., 1944):

"The ancient alchemy of the soil, which transmutes dead organic matter, traces of minerals, and air into the foods which nourish the farmer's and the gardener's plants, is largely carried on by bacterial action. From the soil bacterium, *Bacillus brevis*—'seeded' in laboratory media—we obtain Tyrothricin, a natural bactericidal agent effective against such gram-positive organisms as pneumococci, staphylococci, streptococci, and the diphtheria bacillus."

The medical profession is beginning to get an inkling of what is going on in the soil. It remains for agriculturalists to follow the lead given and to consider that the very same soil organisms that have been adopted by the medical profession with such astounding success are doing marvelous work right in the soil itself by providing a check on the activities of disease-producing, pathogenic organisms. Such disease-producing germs as tetanus and typhoid which are found in the soil may possibly be kept harmless by such molds as the penicillium.

Lady Eve Balfour in *The Living Soil* finds that many soil fungi exist by attacking and consuming microscopic soil animals, most of which are extremely harmful to plants. She states, "The methods by which these fungi trap their prey vary. The majority of the nematode-eating species form loops or bales of mycelium, which also usually excrete an adhesive substance. When an eelworm crawls through these bales, they close round it, holding it captive despite its violent struggles, which may last for as long as two and a half hours."

Dr. Ehrenfreid Pfeiffer in his book *Bio-Dynamic Farming and Gardening* quotes the famous Swiss physiologist, Dr. Emil Abderhalden, in an interesting attitude towards the soil's microbial population as follows:

"In connection with various illnesses of man and animal it has frequently been desirable to trace them back to the method used in fertilizing the food plants. Nothing can yet be said with any certainty, but we must keep in mind the fact that important substances come from soil bacteria, and we must consider whether it is correct to disturb the fine interplay of all the soil organisms by bringing in nitrogen in the form of potassium nitrate and using lime and phosphoric acid, because the development of the various sorts of organisms is thus disturbed and hindered, and on this account difficulties will some day arise."

We cannot go on forever treating the soil as a chemical laboratory and expect to turn out *natural* food. What we are getting is more and more *chemical* food. Instead of eating live matter which can readily be absorbed by the body we are consuming food which is rapidly becoming more and more artificial.

For over a century the practice of eating soil has been indulged in by negroes around Troy, South Carolina. They do not eat just any soil but pick out, by sniffing, a smooth red kind found usually on hilly ground. They have actually eaten hills away. From there the practice has spread to many other rural sections of the south.

Doctors claim that the body receives no nourishment from ingesting such soil, unless it be a small amount of mineral matter. But the negroes claim that it is excellent for stomach disorders and it invariably "quiets de bellyache." If one of them moves to a distant city, his friends will send him earth through the mails. Their children begin to eat it as soon as they quit their mother's breast.

Soil-eaters, of course, are not aware of any scientific reasons for resorting to this practice but practically every such custom that has persisted has some explanation, some justification. It must satisfy some inner urge or feeling that is tied in with their general metabolism. It would be interesting to make studies of the bacteria, fungi, and other soil organisms contained in this particular kind of soil. Perhaps it is rich in penicillium, or there may be some other utterly unknown factor worth investigating. It is probable that these people would not eat soil that had been polluted with strong chemicals.

The healing effect of soil in animal disease is brought to our attention by Sir Albert Howard who advises: "Young pigs are very liable to white scour—a form of enteritis. This can be cured by feeding a turf from a humus-filled soil. A turf from an NPK field has no effect."

Soil pollution is probably one reason why bird-life is greatly reduced on farms where heavy use of chemicals and poison sprays are customary. Many birds like to scratch in the earth and snare insect larvae. One peck in a sour soil that has just been dressed with ammonium sulphate should be sufficient to tell them they are in the wrong place and to fly to greener pastures. The farmer then loses valuable allies in his war against destructive insects.

Department of Agriculture scientists, seeing the amazing powers of soil microbes in curing human ills, are beginning to experiment with them for combatting plant diseases. At the University of Arizona they were able to cure a destructive rot that had been destroying whole groves of giant cacti. Penicillin checked the growth of the bacteria that were causing the trouble and thus saved these scenically valuable plants.

Here again is evidence that in the soil microbial population there is a complicated relationship between the positive and the negative elements. When soil conditions are right, life operates smoothly according to a comprehensive plan. The positive or helpful microbes keep the negative ones in subjection. But when soil routine is disturbed by some disorganizing influence, the bandit, or pathogenic elements enjoy a field-day and neither plants nor crops grow properly.

It is encouraging to note how scientists in many fields are beginning successfully to exploit soil microbes, where up to now chemicals have been depended upon. An alliance should be formed between medicine and agriculture to study them as a whole. Could not a special section in the Department of Agriculture devote its time exclusively to the study of these wonder-working bacteria and fungi? Every new development in the field, both in human as well as plant pathology, should be studied and coordinated with a view to benefitting both agriculture and public health. American farming, on the whole, is still being carried on as if the soil were a sort of mine, without thought being paid to the possible harnessing of our miracle-working soil microbes.

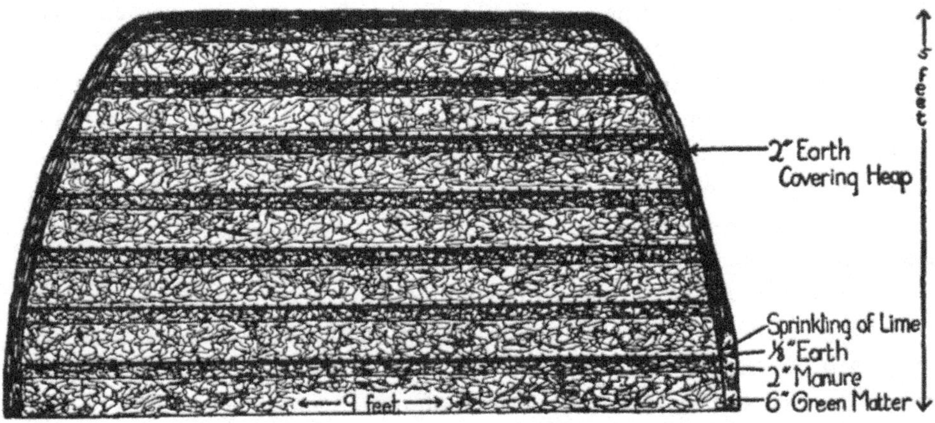

CROSS SECTION OF A WELL-MADE COMPOST HEAP

PART TWO

Compost—Theory and Practice

Repent what's past; avoid what is to come; and do not spread the compost on the weeds to make them ranker.

Hamlet

I

SIR ALBERT HOWARD—FATHER OF SCIENTIFIC COMPOSTING

Sir Albert Howard comes from an English agricultural family long noted for raising fine cattle. He spent his boyhood on a farm that, to a great extent, still followed the agricultural principles of his yeomen ancestors. He inherited conservative farming habits and was privileged to watch a cultural system that paid dividends although it did not push too hard. He therefore looked with suspicion at the modern "scientific" methods that attempted to speed up the clock of agriculture, especially when he observed the train of plant diseases, crop failures, and land exhaustion that went hand in hand with it.

In 1899, at the age of 26, after having received the regular college agricultural training, he was sent as a mycologist to the West Indies to specialize in the diseases of sugar-cane and cacao. He soon came to the conclusion that to engage effectively in research the prober should have a real farm on which to experiment, instead of flower pots, or small plots of earth—the typical set-up provided for research and experiment.

In 1903 he was sent to experiment on insect and fungous diseases of hops at an agricultural college at Wye, Kent, England. Already the germ of his big idea of successful disease control had begun to take shape. Meanwhile he attracted notice by discovering an interesting phenomenon in connection with the pollination of hops—the part played by the male plants in helping the female (hop bearers) to grow and to resist disease.

In 1905 Howard was appointed Imperial Economic Botanist to the Government of India and was stationed at Pusa. He was now in a position to put into practice the ideas that he had developed in the West Indies and at Wye for combatting plant diseases. His program was simply to allow plants to become strong through methods as close to those of Nature as possible. They would thus, he thought, be in an excellent position to *resist* disease and insect infestations. At the outset, he noticed that crops in the regions surrounding Pusa were practically free of both. The natives had never even heard of insecticides or fungicides, nor did they use any chemical or artificial fertilizers, but they always made sure to return to the soil every bit of plant and animal matter that originated on and near their farms.

Howard refused to follow the usual scientific laboratory procedure by placing investigators in separate cubby-holes, and assigning specialized lines of research to each, for he believed that scientific research was far too fragmentized. He obtained 75 acres of land, a grant of money, and the power to test out his ideas on a practical scale. His method of experimenting was to run his farm as a farm. For five years he imitated the methods of Pusa. By 1910 he had learned how to grow crops almost free of disease without chemical fertilizers or sprays. In his mind he established the principle that "*Insects and fungi are not the real cause of plant diseases but only attack unsuitable varieties or crops imperfectly grown.* Their true role is that of censors, pointing out the crops that are improperly nourished. . . . In other words, the pests must be looked upon as Nature's professors of agriculture."

By applying the same principles to oxen he developed in them a remarkable resistance to disease. They were given food raised on land not abused by caustic chemicals. "They were not given inoculations, even though they frequently came in contact with diseased stock. As my small farm-yard at Pusa was only separated by a low hedge from one of the large cattle-sheds on the Pusa estate, in which outbreaks of foot-and-mouth disease often occurred, I have several times seen my oxen rubbing noses with foot-and-mouth cases. Nothing happened. The healthy well-fed animals reacted to this disease exactly as suitable varieties of crops, when properly grown, did to insect and fungous pests—no infection took place." This experiment was duplicated later in many parts of India, always with the same good results.

During the next twenty years he confirmed his experiences at Pusa,

Quetta and at Indore, where he perfected his famous method of making compost. There he had at his disposal a tract of 300 acres and absolute freedom of action. He learned that the best results were obtained and maximum health maintained, if plant and animal matter were composted properly. The actual working out of the "Indore" (or Howard) process took seven years.

The first application of the process was in 1933 at the Kingatori Estate, a coffee plantation of over 200 acres in Kenya, South Africa. In this case about 3½ tons of compost per acre was applied annually. Major Grogan, manager of this plantation, wrote Sir Albert on May 15, 1935, concerning results:

"You will be glad to know that your process is spreading rapidly in these parts and has now become recognized routine practice on most of the well conducted coffee plantations. The cumulative effect of two years on my plantation is wonderful."

In nearby Tanganyika, excellent results were obtained too and the process spread through the efforts of Sir Milsom Rees. Soon work was taken up by Government agricultural workers in South Africa. Captain Timson in Southern Rhodesia eliminated the witchweed trouble in corn and the eelworm disease in potatoes by the use of compost made according to the Howard process.

In 1935 Sir Albert delivered an historic lecture about his process before the Royal Society of Arts. This was translated into Spanish and drew attention all over Central and South America; especially in Costa Rica where it resulted in the adoption of the process on a large scale by many coffee planters.

News of it soon spread to tea-growing estates, and hundreds of them adopted the process. By 1938 about a *million tons* of compost were being made each year on the tea estates of India and Ceylon. In most of these cases chemical fertilizers have been ruled out completely. Experts in tea had long since claimed that after the introduction of chemical fertilizers the quality of the tea had greatly deteriorated. With healthier and better crops grown with compost, there was no need for chemicals.

Howard's process of making compost was soon adopted in a large measure on sugar-cane plantations, by cotton growers, sisal plantations, rice growers, and others, and wherever used, there occurred an improvement in quality, better yields and remarkable freedom from disease. The results in cotton were so interesting that the Agricultural Department in Sind, India, adopted the Howard process and noted that it kept a check on the alkali condition, produced fine healthy plants and increased the fibre yield

The adoption of the Howard process by many municipalities in South Africa is little short of revolutionary. Mr. H. G. Williams, the Ficksburg health inspector, was so convinced of the merits of the process that he installed two pits to transform all sorts of town rubbish, slaughter house refuse, etc., into compost. By 1942 there were eight pits in operation. He calculated that for every 1,000 inhabitants, 25 tons of compost a month could be made, utilizing their wastes, and that the cost of composting would be no greater than the incineration which disposed of the wastes.

J. P. J. Van Vuren, a government scientist, wrote to Sir Albert on May 4, 1943, stating, "We have so far over forty municipalities that have adopted the scheme and many more are soon to follow." There are dozens of towns in Malay that are—or were prior to Japanese occupation—converting all manner of town wastes. In 1943 the Government of India adopted the composting plan for all urban refuse in the entire country. It will, of course, take many years to put it completely into practice.

Dr. J. W. Scharff, former health officer at Singapore, put Howard's process into effect there and by September, 1939, all the incinerators in the outlying sections were shut down. Dr. Scharff escaped before the Japanese arrived, and subsequently engaged in composting activities around military camps in England.

It was due to the efforts of the *Farmers' Weekly* published at Bloemfontein, South Africa, that Howard's discoveries spread through that country. They have published dozens of his articles.

In 1942 the Humic Compost Club was formed in New Zealand. It publishes the interesting *Compost Club Magazine* and has several thousand members. From its articles it can be seen that the organic idea is spreading rapidly in that country. In 1945 a similar organization was formed in Australia.

The *News Letter of Compost* has become the English sounding-board for Sir Albert Howard, and it is unequivocally against the use of artificial fertilizers. It deals exclusively with compost farming and gardening and their relation to health. The Cheshire Panel sponsors a garden competition each year. There are two conditions: only compost must be used, and all artificial fertilizers must be excluded.

Howard's ideas are spreading rapidly in England, where his process today is in wide use. In January, 1944, a demonstration of a new piece of equipment which automatically turns compost heaps was attended by 500 farmers. Over 100 machines were sold on the spot. The spread of the method in England has reached such proportions that two long debates were held in the House of Lords in 1943, and again in 1944, for the purpose of urging the Government to appoint a Royal Commission to investigate its advantages.

In the United States the work of Howard is being made known and many American gardeners, as a result, are discarding the use of artificial fertilizers and becoming adept in making compost by the Howard process. Even with the present hand methods of making compost, the exclusive organic method can compete favorably in cost with the chemical system. When the compost process becomes mechanized thousands of our farmers will be quick to see its advantages and the system will spread rapidly.

It should seem unnecessary to emphasize that any farmer or gardener can benefit greatly by reading his great books, *An Agricultural Testament,* which details his long and profitable experience with farming in several lands and climates, and *The Soil and Health,* which continues the story and brings it up to date.

2
HOW TO MAKE A COMPOST HEAP

Compost is made so that raw substances may not go into the land. In order not to interfere with growing processes all materials intended as organic fertilizers should be pre-digested, or "finished" before being placed in the soil. Otherwise, soil bacteria will finish the process in the ground at the expense of the new crop. Compost is so made that when it is plowed in it is ready to go to work immediately in the interests of the current planting. It is not good practice to plow under raw, fresh manure or a heavy mulch of straw or hay, but when these are properly combined and mellowed in the compost heap, spectacular farming or gardening results will accrue.

The instructions which follow represent the *Howard Method,* known also as the *Indore Process.* This consists in mixing vegetable and animal wastes with earth and water. Since vegetable matter, decomposing alone, is always acid, it is not practical for growing vegetables and commercial crops. To counteract this acidity, we must mix our vegetable and animal wastes and combine them with a base, such as earth mixed with lime or wood ashes. The materials must be handled in such a way that the micro-organisms which break down the raw matter can do their work in the most efficient manner, and also to prevent certain conditions from arising that give birth to outlaw organisms which putrefy, rather than compost, organic matter.

Location. Pick a location that is shaded, if possible. Where a sunny spot is unavoidable, it is wise to cover the heaps with a straw or hay mulch, or mats. The ideal site is one that has a wall to the north of the heaps and windbreaks to the east and west. A strong wind will stop fermentation on the sides of the heap most exposed. Make your arrangement of heaps orderly and as neat as possible with room to walk between and around them. There should be sufficient room for tractor or truck if they are to be used.

Pick a high site which drains well, otherwise rains may wash into the heaps. If you are in a dry section arrange for a convenient supply of water. Leave space for piling heaps of green matter—leaves, grass clippings, weeds, etc., so that they may wither some before being placed in the heap.

Some farmers make their heaps in the particular fields where they are to be spread figuring that labor in hauling will be saved. Such is not the case. In fact, the reverse is true. Due to the shrinkage that takes place in composting there is less to haul of the final product. If you have a composting center near the barn you can gradually arrange for ideal conditions there such as windbreaks, water supply, and other factors. Use the same site over and over again so that the ground underneath becomes progressively richer with bacteria. When a new heap is built these micro-organisms will invade it and thus hasten the breaking-down process. For this reason it is advisable to leave some of the old compost at the site to act as a culture for the raw heap. Some farmers sprinkle old compost in the new heap itself as it progresses, to accelerate bacterial action.

A compost heap should not be directly under a tree because its roots will feed on the ingredients of the heap and valuable nutrients will be lost. Second, trees spread their roots into it and the pile becomes run through with a system of branching roots and rootlets. The only solution where no other space is to be had is to build a bed of branches about seven or eight inches thick as a base for the heap. This, of course, will not give the valuable contact with the ground, but it is the lesser of two evils and while other means may have to be used to encourage bacterial activity, the branches help in aerating the heap from the bottom.

Beginning. First remove the grass sod from the spot where the heap is to be, so that the bacteria in the soil may have immediate contact with the materials. Grass sod forms an acid layer which discourages even earthworms from entering the heap. Some persons dig out five or six inches of earth to make a very shallow pit. Others make pits about 30 inches deep. This is usually advisable in cold and windy climates and prevents the heap from drying out. Where a heap *is* made directly on grass, the latter will be killed off so that when the next heap is made on the same site there will be an earth base. It is best, however, to remove the grass first. Some persons make the mistake of building a pit with a concrete floor. This will not give the best results as bacteria do not thrive on new concrete. Another objection to concrete is that it draws too much heat away.

The Green Matter. In this category are included all kinds of vegetable matter such as leaves, weeds, grass clippings, hedge-trimmings, sea-weed, spoiled hay and straw, kitchen vegetable wastes, chaff, prunings, garden residues, sawdust (not over 5% of the heap), tea leaves, coffee grounds, and any other kind of plant material you can get. In many cases you will find neighboring land or roadsides growing wild in weeds. By cutting them down you not only obtain valuable material, but prevent the weeds from going to seed and spreading to your land. Much plant material can be obtained in neighboring cities. Brewery wastes and grocery store residues are two examples.

On farms it is a good practice to set apart a certain time of day to go out with a truck for a load of weeds, leaves, and other material. It is surprising how such daily loads will soon accumulate a vast amount of vegetable matter which may be permitted to wither before being used. This is material that ordinarily goes to waste and disfigures the farm besides; but it represents dollars and cents in fertilizer value. Green matter contains all of the elements needed in the soil. Besides nitrogen, phosphorus and potash, it contains calcium, iron, magnesium and many others including the rare trace elements such as boron and fluorine. By itself, however, it is insufficient as it lacks the activating substances which animal manures contain. Sir Albert Howard in his book discusses this:

> "In Ceylon particularly, attempts have been made to prepare humus without animal wastes. The results have not fulfilled expectations. The breaking down of such resistant material as the leaves and prunings of tea is then unsatisfactory; the organisms which synthesize humus are not properly fed: the residue of these organisms which form an important part of the final humus lack the contributions of the animal. No one has yet succeeded in establishing an efficient and permanent system of agriculture without livestock. There is no reason therefore to suppose that the tea industry will prove an exception to what, after all, is a rule in Nature."

The vegetable wastes should be well-mixed and then thoroughly moistened. Do not worry about diseased plant material. It may safely be included in the heap, provided it gets into the center some time where the heat can break it down thoroughly. To test this out a large tomato grower in the south of England made compost out of an entire diseased crop of tomato plants and used it to grow a new crop. The new plants were free of the disease. In the same way most of the weed seeds will be killed off, provided in turning the heaps, as will be explained, every part of the material gets its chance to occupy the *inside* part of the heap where the heating and fermenting action is strongest.

It is extremely important that a wide variety of plant material be used in making compost because no two species have the same make-up. As Dr. Ehrenfried Pfeiffer has pointed out, "the more varied the compost, the more certain are its dynamic effects. Some plants have importance in directing the fermentation of organic matter; others are 'specialists' in producing a substance which the soil lacks, thus contributing to soil improvement. . . . An imitation of this process can be accomplished in a practical way by making compost of *everything:* the greater the variety of plants used in making compost the richer and more useful it is in its nutritive potentialities." (*Bio-Dynamic Farming and Gardening.*)

Pfeiffer then cites a few examples: Foxglove likes to store up iron, calcium, silicic acid and manganese. The dandelion is extremely rich in calcium and silica. The wild sugar-beet is a small "pharmacy" consisting of goodly amounts of sodium, lithium, manganese, titanium, vanadium, strontium, caesium, copper and rubidium. It grows at the seashore and along deeply indented bays and fiords. Yarrow is rich in potassium, calcium and silicic acid.

Every compost heap should include some leaves because as a rule they contain very valuable elements. It is not unusual for the roots of a tree to descend to levels of 20 feet or more, where the soil hasn't been "mined" by intensive cropping methods. In cases where surface soil is exhausted the deep subsoil is apt to be still rich in essential minerals. The roots probe into these lower levels and bring up food to the leaves.

Sir Albert Howard describes a situation much to the point at Pusa in India. It is a recognized fact in agriculture that to grow good crops the soil must be amply supplied with phosphorus. At Pusa the soil was found to be practically devoid of this element. Yet the finest crops had been raised there without the aid of chemical phosphates—crops so fine that the seeds are exported from Pusa to all parts of India:

"The living roots comb the upper 20 feet of soil for such minerals as phosphates and potash which are used in the green leaves. These leaves in due course are converted into humus and help enrich the surface soil. This explains why the soils of North Bihar, although very low in total and available phosphates, are so exceedingly fertile and yield heavy crops without any addition of mineral manures."

It is advisable to gather leaves separately and allow them to weather for a few months—the longer the better—before placing them in the heaps. In the meantime they should be watered and stirred from time to time. When adding them to the heap be careful not to put them on too thickly. If they cake it will stop aeration, and decay processes will be retarded. Probably not more than 20% of the green matter should be leaves.

Stubborn Matter. Certain kinds of green matter resist and will not break down as quickly as other materials. Such items as corn stalks should first be finely shredded, and corn cobs, if used, should be ground up. They both may be used to better advantage as mulches and, when well-rotted, plowed under.

Apple pomace, the substance of apples remaining after the juice has been squeezed out, is exceptionally stubborn. While it will impart a pleasant aroma to your composting center, it does not break down for long periods. Pine needles are another slow actor. It is best to put them in separate, shallow heaps, keeping them well watered, and well stirred.

Woody material, twigs, branches and the like, should be thoroughly crushed before going into the heap. They can first be put on ground where trucks and other heavy equipment go over them. Cotton stalks must be given the same treatment. Sugar-cane trash is exceptionally difficult. It has leaves that seem to be armor-plated. They must first be allowed to wither and kept well-saturated with water. Mixing in molasses seems to help the fermentation of these leaves.

At times you hear people say not to use orange or grapefruit rinds but these may be safely used, and used plentifully. Palm-tree fronds are one of the most stubborn of materials. The city of Miami, Florida, has a special machine that grinds them very fine before placing them in compost heaps.

Do not use too much paper. A small amount sometimes may be unavoidable, and will not be harmful, but because paper is treated with bleaching and other chemicals, and the ink contains poisons, it has nothing to offer to compost.

You will soon learn to examine materials and appraise their compostable qualities by experience; sometimes by sad experience.

Construction. The size of the heaps will depend on whether they are for farming or for gardening, and on the size of the acreage or plot used. Make your heaps as large as possible, within reason, of course. The smaller the heap the quicker will heavy rains slow down the fermentation. One large heap always does better than a number of small ones. There is no rigid rule as to size except that if you make heaps too narrow they will tend to dry out, and if you make them too wide air will not be able to penetrate to the inside. The minimum width, therefore, should be 5 feet and the maximum about 12 feet. The length can be almost any size. Heaps have been made 30 to 40 feet long. A good size for the average small garden is about 8 feet wide by 12 feet long.

Start by putting a layer of green material about six inches high on the ground, to the width and length that you have chosen for your heap. This is then followed by a layer, about 2 or 3 inches thick, either of manure or a recommended substitute such as fish or chicken cuttings, or animal matter taken from your kitchen waste.

You then take earth, good, rich top-soil, if possible, and mix it lightly with ground limestone or wood ashes. Put this combination over them to a depth of not over ⅛th of an inch. Be sure that it is agricultural limestone, not the strong, builder's "lime" called gypsum. Earth saturated with ground limestone acts as a base to neutralize acidity. The earth also contains bacteria which will contribute to the decomposition processes. For many years it was believed that lime should never be placed next to manure as it would liberate much of the nitrogen of the latter. Sir Albert Howard, however, found that there was no harm in it. Recently this opinion was backed up by experimental work both at the New Jersey and Vermont Experimental Stations. They found that Howard's experience is true; that lime next to manure actually prevents nitrogen from being dissipated into the atmosphere. Many farmers and even some agronomists will be surprised to hear this.

To review, you have six inches of green matter, two inches of manure and a light sprinkling of earth saturated with lime or wood ashes. Then you repeat the process over again: green matter, manure and earth—green matter, manure and earth until you have reached a height of five feet. Keep watering the heap as you make it, but don't let it get too soggy. It should have the consistency of a wet sponge. The right amount of moisture is important and will come with experience.

As you make the heap do not tamp it down but leave it soft and fluffy so that air can get right into the center. For this reason do not walk on top of the heap any more than you have to.

As you construct the heap, let it taper up, so that if it is 8 feet wide at the bottom it narrows to about 6 feet at the top. Also scoop out the center of the top, saucer fashion, so that the top of the heap will retain as much of the water as possible when it rains. The last layer of manure should be a little thicker than the others. The bio-dynamic people urge that the last layer of earth also be much heavier and cover the sides as well as the top; but this may slow the process and is bound to impede aeration.

If the weather is wet, you need not water the heap, but it is essential to do so during dry spells. Good results will be obtained if you cover the heap, top and sides, with a straw mulch about six inches thick to prevent the top of the heap drying out when the weather is hot and dry. Also, it allows sprayed water to reach the heap more uniformly. Without this mulch be careful to use a fine spray of water on the heap. If you can obtain run-off rain water or water from a pond for spraying, so much the better.

When the heap is complete, make three or four equally spaced holes in the top to allow air to get to the different parts. These holes should be five or six inches in diameter and run from the top to the bottom of the heap in the center of the width of the top. Holes can be made by a crowbar, but a better method is to insert pipes as you build the heap and to pull them out when it is completed. We use this method on our farm. It produces big, round holes clear down to the bottom.

Turning the Heaps. Three weeks after the heap is built it is turned so that what was on the outside goes into the inside. In this manner every bit of material has a chance to undergo the heating, fermenting, decaying action of the inside, where the bacteria do their best work. Five weeks after the first turn, it is turned again. It is best to use a five-prong pitchfork to make these turns. Four weeks later, or exactly three months after the heap was made, it is ready. It does no harm if it stands a little while after the three month period. In fact in some cases it may take four or five months or even longer for complete decay, due chiefly to climate. Naturally, in a hot or warm climate decay is faster. When the compost is completed it should be applied to the land as soon as possible, or, if not used for a while, covered with a canvas or mulch.

Chemical Activators. Practically every Government agricultural experiment station has issued a bulletin describing the production of artificial manure, or compost. But unfortunately they recommend chemical activators, in the form of the worst chemical offenders such as ammonium sulphate (which is instant death to earthworms), cyanamid, nitrate of soda, urea, superphosphate, etc. All of the arguments against the use of chemicals direct on the land apply equally well to their use in the compost heap. Stress is laid, in their use, on the mere obtainance of big crop yields without considering quality, disease resistance, or other important factors. But any novice can tell at a glance by comparing the two finished composts which is of better quality, the one having the odor of fragrant woodsy earth, the other a noxious smell and coarser appearance.

3

THE COMPOST PROCESS

Making compost is an art rather than a science. To go about it mechanically, merely following rules, not only will not yield the best results, but the work will not be as enjoyable. We must understand first of all that composting is a process in which bacteria, fungi and other microscopic organisms assume the main role. While there are chemical changes in the heap, it is essentially a biological process.

The first thing to bear in mind is that the microbes playing a part in the process must be fed properly, on food furnished by the manure and green matter of which the heap is built. Manure especially supplies nitrogen for the bacteria to feed on while breaking down cellulose. They need phosphorus and potash too, but in much smaller quantities, possibly only 10% as much. These they get from both the manure and the green matter.

Next there must be proper working conditions. As soon as the heap is made, a very strong fermentation commences, and the temperature goes up to about 160° fahrenheit. It stays at about this point for quite a while and then gradually goes down to 90°. The question is sometimes asked, "If a temperature of 150° or more will kill bacteria, how can decomposition be brought about with bacteria playing such an important part in it?" The answer is that as the temperature varies, different kinds of microbes come into action. There are certain kinds of bacteria called *thermophilic* which can withstand very high temperatures and many groups of actinomycetes develop when the temperature goes above 150°. At certain stages, especially at the beginning, the fungi take a more prominent share in the work. After the second turn bacteria become the more active because the temperature has declined and comes into their range. These bacteria are largely anaerobic, that is, they do not need air. For this reason air holes need not be made in the heap after the second turn.

It is remarkable how versatile nature can be in the processes of the compost heap. At various stages she uses different forms of bacteria and fungi. But very little is known about the exact function of each and much additional scientific study is needed. Professor Selman A. Waksman of Rutgers University is a world-acknowledged authority in this field and he has engaged in extensive researches, but mainly on composts made with stable manure only, or with manure and chemical activators, or with sewage.

A good deal of thought has been given to methods of stimulating the biologic life of the compost heap. One scheme is to douse the heap from time to time with water rich in soil organisms, which is obtained by soaking rich garden soil. Two pounds of soil is shaken up in four gallons of water and then permitted to stand for an hour. This idea will no doubt prove very practical for small gardens. The average farmer, however, will not have time for it.

There are cultures of bacteria on the market which it is claimed will work very effectively and speed up the decomposition in the compost heap. We have tried them many times but have never been able to get the bacteria to work. Professor Waksman in a letter to the writer on this subject, reports, "Studies made on the artificial inoculation of composts with either pure cultures of bacteria or with mixed cultures of different groups of micro-organisms have never given us very satisfactory results, to justify any claims made by various people (such claims have usually been based upon insufficient experimental evidence) that composts benefit from inoculation." The microbial processes in the heap with the many changes of temperature are so complicated, and so many of them come in and out of action at the various changes, that the insertion of a culture would seem to be an unnecessary contribution.

Average raw stable manure contains over 20% bacteria, from the animal's digestive system where they were used to break down food. If fresh manure is used in the heap, that should be a sufficient start. If you feel that you want to make an additional contribution you might add the inoculated water described above. You can also strew some old, finished compost in a new heap as it is being made.

If a compost heap is made correctly it should begin to heat up in a few days. You can stick a metal rod into the heap and then check to see if it is warm when you pull it out. Another sign is vapor coming out of the aeration holes. This is not a danger sign as some suppose. If you want to be technical about it you can mount a thermometer at the end of a stick and take daily readings.

Within a few days from the start, the temperature should reach about 160° and stay at this point for a few weeks. You can imagine the effect on weed seeds of such intense heat combined with moisture.

During the first eight weeks, air is needed in the heap to insure that decomposition is effected by fermentation and not putrefaction. That is why the materials must be arranged loosely, although as stated they should not be too loose. It is extremely important to aerate the heap during the first eight weeks, so that oxygen can penetrate easily and carbon dioxide escape.

In making aeration holes some people take material like rye straw, corn stalks or broom sedge and make it into bundles about four inches or more in diameter, tied loosely so as not to be too compact, and stand one of these in each place a ventilation hole is desired. Those who prefer this method claim that holes made by a crowbar cave in.

If the heap doesn't sink in a few weeks, it is a sign of poor aeration and that action has slowed down. Another bad sign is an odor of ammonia emanating from the heap, which may mean that it is too tightly packed. A too wet condition will cause the same thing. It is best in such cases to make the heap over again. Do not allow weeds or grass to grow on compost heaps. Their roots cut off the supply of air.

The failure of a heap to settle may also be an indication that it is too dry. If you are checking temperature carefully and find that it goes over 150° the heap should be given a thorough wetting. If the heap is allowed to become too dry, fermentation gradually ceases and the compost takes much longer to mature, with a loss of many valuable nutrients. When heaps must be some distance from the water supply, you are, of course, at the mercy of the climate. The presence of ants or wood lice is an indication of dryness.

Occasionally, people ask whether dish-water is good for compost heaps. The answer is "No." Such water too often contains strong acids and greasy matter which impede the action.

Compost water is excellent for watering seedlings and for other uses in fine gardening. It is made by placing water and finished compost in a barrel and then using such water. The compost can be used several times. It can be used longer if occasionally a little new compost is added to strengthen the solution.

There is very little that can be said about turning the heaps except that the dates for turning be adhered to as closely as practicable. The first comes in three weeks, the second, five weeks after the first. These dates need not be followed to the minute, and may vary slightly depending on the weather or individual conditions. It is advantageous if the turn is made on a rainy day. It is advisable to have the dates noted on something that is stuck into the heap, especially if you have many of them. In such cases, it may also be advisable to keep a reminder system on your calendar.

4

LIME, WOOD-ASH AND ROCKS

Lime is added to the compost heap in order to maintain the proper condition of alkalinity; so that the soil microbes can perform their function of decomposing matter. It also helps to fix the nitrogen by converting volatile compounds into stable compounds.

Lime comes chiefly from a particular kind of rock-limestone. It is one of our most common rocks. All soils originally were formed from rock and the process of soil formation is still going on. Soils formed from limestone are considered by many about the best for growing crops, although there are cases of poor soils on top of limestone formations.

Ground limestone is also best for use on the land as well as in the compost heap. The finer it is ground the better. Limestone is not chemically pure carbonate of calcium, but often contains quite a percentage of carbonate of magnesium. Magnesium is a very useful plant food and this "impurity" really improves the limestone. When limestone is "burned" it turns into calcium oxide or quick-lime which is much too caustic in its action on micro-organisms. F. H. Billington in his book, *Compost,* advises the use of quick-lime, however, for heaps made up exclusively of vegetable matter.

When water is added to quick-lime it changes to slaked or hydrated lime. Hydrated lime should not be used in the compost heap unless ground limestone is not available, and in such event it is still better to use wood ashes.

There has been much confusion as to which form of lime to use. Experimental work shows conclusively that ground limestone is first, hydrated lime next, and quick-lime worst.

Sir Albert Howard, in an amusing letter, gives his opinion on the best form of lime to use: "The lime question reminds me of the changing fashions in ladies' dresses and hats! When I was a student the great thing was ground quick-lime (CaO). Farmers always used slaked lime (mostly $Ca(OH)_2$ and some carbonate). In olden days chalk ($CaCo_3$) used to be common. Now the fashion is to use limestone meal (impure $CaCo_3$). Personally I think the carbonate (ground limestone) should be used in composting, as it is not so fierce as the hydrate."

The popular use of quick-lime in olden days in agriculture was due to the fact that they did not have the machinery necessary to grind the limestone as fine as we can today. Many farmers had limestone kilns on their land where limestone was burned for this purpose. On Pennsylvania farms today many of these can still be seen although they are rarely used. Such is habit, however, that although the need for it is gone, many farmers still swear by quick-lime.

There is a widespread fallacy, as we have seen, in connection with the application of lime to manure. General advice to the farmer has always been not to place lime next to manure. One company which sells a chemical activator to hasten the decay of the compost heap has consistently warned about this. Even the U. S. Department of Agriculture has held this view. *The Garden Dictionary* says, "Do not add lime in any form to any manure heap, nor to a compost pile. It does more damage there than anywhere else for it greatly increases the loss of nitrogen."

This idea must have become established when the use of quicklime was popular. *Bulletin 707* of the New Jersey Agricultural Experiment Station gives the results of experiments which indicated that not only did the use of hydrated lime with manure save the nitrogen, but it actually fixed a great part of the nitrogen that might otherwise be lost. The same thing happens in a compost heap.

For 40 years, Howard used ground limestone in compost heaps successfully. The fact that it worked over so long a period is all that is necessary. Professor Emerson, of Ames College, Iowa, who was the Chief of the Soil Survey Division of the U. S. Department of Agriculture, has stated that limestone used either with plant or animal manure often increases its value as much as 400%.

Gypsum (hydrated calcium sulphate) is a form of ground stone that contains sulphur and is definitely not recommended either in the compost heap or directly on the land. It has the opposite effect of lime, as it makes the soil acid, not alkaline, due to the acid sulphur radical of the gypsum accumulating in the soil. In the olden days it was called land-plaster and used extensively. The sulphur in gypsum gives rise to certain sulphur-reducing bacteria which are known to kill off important and beneficial soil fungi.

Wood Ashes

Wood ashes are better than lime in the compost heap because they are quicker-acting, possibly because they supplement the bacterial needs in many more ways than lime does. Hardwood ashes are better than those of soft wood because they contain more potash and because of their density or weight. The most generally used classification places the deciduous trees into the hard wood class: oaks, maples, the fruit trees, etc. The conifers are referred to as soft woods and include the pines, spruces, hemlocks, firs, cedars, etc.

At one time hardwood ashes were the chief source of potassium and were extensively used in farming. Those were the days before modern conveniences found their way onto the farm. Unleached hardwood ashes contain about six per cent potash, about two per cent phosphoric acid and thirty per cent lime. If left out in the rain and leached, the potash content may do down to as low as one or two per cent. Wood ashes contain small amounts of magnesium and other elements, but no nitrogen.

If too much of this substance is applied to the soil, especially an alkaline soil, the plants may have pale, undernourished looking tops, and the soil may become hard and lumpy.

Paper ashes should not be used in appreciable quantities, as they contain strong acids and other harsh substances used in the manufacture of paper which are not good for growing crops.

Lime as a General Fertilizer

The benefits to be derived from liming soils were known to the ancient Egyptians. Lime is not a strong chemical and therefore may be used on the land with safety. Most lime is organic in origin. Limestone beds are mostly ancient sedimentary marine deposits, made up of fossilized shells, corals, and skeletons of marine animals.

Although ground limestone is used in the compost heap it may also be used directly on the land. It may even be used where a test already shows an alkaline condition, because the commercial product contains some magnesium and other elements. There are many limestone soils which contain an underlying limestone formation, yet are starved for lime or calcium due to unwise farming practices. Such soils need limestone.

The use of ground limestone increases the number of bacteria in a soil unless too much is used. Ground limestone seems to improve the soil structure also and its application increases crop yields. In an experiment at the Georgia Agricultural Experiment Station where limestone was added (400 pounds per acre) to the chemical fertilizers, the pea and vetch crop increased from 506 pounds per acre to 4,092 pounds. O. B. Price in the Michigan Agricultural Experiment Station *Quarterly Bulletin* for 1923 described a 500% increase in calcium and a 300% increase in the phosphorus content of alfalfa where two tons of ground limestone had been applied to the acre. Lime is a helper in moving other elements into the roots besides going in itself, phosphorus, for one.

Ground limestone may be applied almost any time. One method it to apply it every time corn is planted in the rotation. Lime seems to give marvelous results with all legumes, such as clovers, alfalfa, peas, etc. On sod fields it may be spread even in the winter. Do not use it in the case of certain vegetables such as potatoes or for blueberries.

We have the statement of Dr. William A. Albrecht, Chairman, Department of Soils, Missouri College of Agriculture, that from personal observations cattle—left to their own devices—would turn down good appearing pasture in near-by acres and travel a long distance to eat the crops on plots that were most abundant in minerals, particularly lime.

In England a great deal of chalk is used. Chalk is built up of finely divided fossilized shell particles. It has twice the bulk of lime but much less value, and is, therefore, a poor substitute. Marble is metamorposed (changed) limestone, usually very pure, and crystallized by pressure and heat. In some parts of the country marble dust is sold for use instead of lime. As long as it is merely pulverized and not hydrated it would be as satisfactory as chalk.

Limestone may be applied at the rate of from 500 pounds to three or four tons to the acre. For small gardens about 50 pounds to 1,000 square feet should be sufficient. If hydrated lime is used the amount should be reduced by about 30%. For garden use place the lime in a burlap bag and shake it over the surface. It can then be worked into the soil. It can be applied in the fall, winter, or spring.

Ground-up Rocks as Fertilizer

There is a small group of agriculturists who believe that farming can be carried on with ground-up rocks, usually called stone-meal, without the use of animal manures. They call themselves *clean culturists* and their system is known as "clean culture." This theory was first offered by Julius Hensel at Christiana, Norway, in 1885. He was opposed to both manure and commercial fertilizers. Some scientists sided with Hensel, others sneered, and a heated controversy arose in the press. A few factories were erected to manufacture this stone-meal and in 1892 an exhibition was held in Leipzig, Germany, at which were shown crops grown with this dust.

In 1924 Jack Gaerity, an American farmer, resurrected the work of Hensel and wrote a booklet based on it, *Bread and Roses from Stones.* In it he says, "Stone dust has been successfully used in England, but it faces the concrete wall of academic opposition there, too. For several years, I am informed, a stone dust fertilizer has been on sale, under the trade name of *Fluora.* The British Board of Agriculture, dominated by academic scientists, will not permit it to be sold as fertilizer, so it is distributed as 'soil dressing.' " Then he quotes J. A. Minchin, of Henfield, Sussex, as stating: "I have, even this dry summer (1919), grown most splendid salads with the aid of *Fluora.* My winter green stuff, Brussels sprouts, etc., are also looking remarkably well and strong." In further support he cites a German farmer, B. Wettengel, in the *Wiesbadener General Anzeiger* of July 1, 1893, as "using stone dust fertilizer in my garden and fields for five years. The results have always been satisfactory in every respect, for the soil becomes better every year by using this fertilizer. Especially this year during the extraordinary drought, the excellent effects of stone dust fully manifested themselves. The flowers as well as the different vegetables developed so magnificently that every person who passed my garden stopped and admired the great growth, especially of the Kohlrabi."

Hundreds of other British and German users might be similarly quoted in praise of it, according to Gaerity. The writer has come across a few persons who are following Julius Hensel's teachings. They usually refer to animal manures as "filthy stuff." However, they do use vegetable composts. Some of them use such composts exclusively as they are unable to get the stone-meal. But they will not use animal manure in any form. None of these persons has carried on long enough in one location really to test out the idea.

No acceptable substitute for animal matter has as yet been devised, nor is there anything as simple to handle without fear of disastrous consequences due to burning, overdoses, and all the other complications. I feel quite certain that the fertilizer formula of the future will consist of plant matter, animal matter and finely ground up rocks. This is Nature's way.

Dr. William Crocker, a plant physiologist, writing in *The Manufacturer's Record* of November 27, 1919 stated that, "a balanced ration of natural rock fertilizers is possible." Professor Albrecht in an article in *The Land* recently very aptly said, "What is soil, after all? It is a temporary rest-stop while the rock is on its way to the sea." It is an elementary fact that soils have been formed from rock, but why wait for the slow mouldering process of rock disintegration? There are for example many phosphate rock deposits in this country which are being sold daily in a ground-up form as fine as flour. But this is not satisfactory to some of the college agriculturists. It is still too slow acting. It is too mild. What they want is something that will give the earth a shot-in-the-arm and produce results next week, regardless of the evil residual consequences. So to the ground-up phosphate rock they add highly corrosive sulphuric acid and produce the much-used superphosphate. Raw phosphate rock ground up fine may be safely used, but not its cousin, superphosphate. Phosphate rock may be slow-acting; still next year does have a habit of rolling around.

In the United States there is an enormous amount of potentially valuable fertilizer rock. A survey might be made to ascertain its special qualities, and tests made of its effects on different soils. Some undoubtedly have caustic properties and could be eliminated at the outset.

In connection with ground rock, the story of the Hunzas of India is interesting. The *Wheel of Health* by Dr. G. T. Wrench of London is a saga of the Hunzas, whom he called the healthiest people in the world. The Hunzas, who live on the northwest frontier of Kashmir, have tremendous endurance and are extremely good-tempered and cheerful. They think nothing of covering 60 miles at one stretch, climbing over mountains on the way, doing their business and then returning. The Royal Geographical Society in 1928, in describing them, said, "The Hunza men were with us two months, continuously on the move, over what is probably some of the worst country in the world for laden men. Always ready to turn their hand to anything, they were the most cheerful and willing set of men with whom we have ever traveled."

Mildred Jensen Loomis in an article in *Organic Gardening* (May, 1943) writes, "They (the Hunzas) return everything they can to the soil—cattle manure, vegetable parts and pieces not used for stock food, fallen leaves, human sewage from privies after six months, silt from recesses in their irrigating channels, wood ashes from fires; all these are mixed together and made into a compost. This, along with alkaline earth from the hills, they spread carefully on their vegetable fields."

Anent use by the Hunzas of ground-up rock, Sir Albert Howard in his *Agricultural Testament* says:

"There is one point about the Hunza agriculture which needs further investigation. The staircase cultivation of these hillmen receives annual dressings of fresh rock-powder, produced by the grinding effect of the glacier ice on the rocks and carried to the fields in the irrigation water. Is there any benefit conferred on the soil and on the plant by these annual additions of finely divided materials? We do not know the composition of this silt. If it contains finely divided limestone its value is obvious. If it is made up for the most part of crushed silicates, its possible significance awaits investigation. Do the mineral residues in the soil need renewal as humus does? If so, then Nature has provided us with an Experiment Station ready-made and with results that cannot be neglected. Perhaps in the years to come, some heaven-sent investigator of the Charles Darwin type will go thoroughly into this Hunza question on the spot, and will set out clearly all the factors on which their agriculture and their marvelous health depend."

Rocks—Parents of Soil

For millions of years the earth's crust has been "weathering"—its rock structure decomposing through the action of heat and cold, winds, rains, fogs, glacier movement, climatic change, and the life and death cycles of plants and animals.

The soil that is thus being formed represents only a thin skin, compared to the rock mass which is still weathering and forming soil underground. This continuing addition of soil takes the place of some of the surface soils carried off by erosion. The underground mass of rock is the soil parent and plays an important role in the action and history of the soil's cultivation, although too little is as yet known about it.

It has been discovered that the well-being of cattle and human beings is closely related to the kind of soil which grows their food, and this, in turn, is influenced by the kind of rock out of which the soil is formed. Miscellaneous Publication No. 369 of the U. S. Dept. of Agriculture, *The Mineral Composition of Crops* points to the critical need of further investigation in this field.

According to Publication No. 369 the following rock formations are apt to weather into inferior soils: Sandstones, Aplites, Granites, Pierre shale, Cretacious rocks, Air-borne pumice (an acid lava) and Volcanic.

The better soils are on Limestone, Basalt, Dolerite, Diorite and Gabbro formations. With regard to Gneiss soils the results are sometimes good and at other times unfavorable. There are so many cases given that it is difficult to know where to begin to summarize. In 1931 in Florida a "salt sick" disease, an anemic condition of cattle, was found to occur in practically all parts of the state but only on light, sandy soils. No such sickness occurred on clay soils. In New Zealand in 1932 it was found that bush sickness was found mostly over granitic formations. A bone disease of livestock was discovered in Australia in 1895 in cattle fed largely from granite soils, where there was a deficiency of lime.

In 1920 there was a deficiency disease called "creeping sickness" in southern Alabama along the coastal plains which was found only among people who lived on sandy soils low in lime. In 1924 in South Texas there was noticed a fatal disease in cattle called "loin disease" or "down-in-the-back," in a section where the soil was notably lime-deficient. A nutritional anemia in Florida was discovered in 1934 in the flatwoods country which affected 96 per cent of the children in that section. The soil, known as Leon, is thin and sandy, and low in lime content. Six miles away where the soil overlays hard rock phosphate only 3 per cent of the children had this anemic condition.

These are only a very few of the cases mentioned in Publication No. 369, but they indicate the need for a national survey of soil origins in connection with human disease.

Deaf Smith County, Texas, figured in the newspapers a few years ago as "the county without a toothache." The soil in Deaf Smith derived from rocks that are rich in lime and phosporous and contains some fluorine which is important in the making of bone and teeth. Even when strangers came to Deaf Smith County, the cavities in their teeth seem to glaze over and progress no further. Farmers bring spindly cattle from Mexico there and in a short time make fine big-boned animals out of them. What kind of rock structure you have under your soil is therefore important. It is a large factor in determining the nature of good soil.

The healthy Hunzas have been discussed. Right across the river from them live the Nagyrs, who are of the same race, but whose health record is entirely different. The Nagyrs are subject to goitre and many other diseases to which the Hunzas seem to be immune. The Nagyrs are of smaller stature and dour looking as compared to the charming and agreeable Hunzas. From what I can gather, one of the reasons for these differences is that the Hunzas have a much more desirable geologic formation under the soils on their side of the river.

Roger Babson, the financial expert, in his syndicated newspaper column of Feb. 4, 1944, wrote: "Some years ago when I was in Switzerland making a study for the *New York Times*, a Swiss scientist said to me: 'Europe will always have trouble with those Prussians. There is something in the soil of Prussia which grows people with that fighting instinct. European peace is unalterably linked up with the minerals of the soil.' " This bears looking into. Perhaps the new world organization to maintain peace might study the subject thoroughly: it may have a great deal to offer in explaining national behavior.

Later on in the section, *Is Our Health Related to the Soil?* (p. 150) we discuss the problem of deficiencies in certain soils. Many sections of this country have unfavorable geological formations. This then, becomes one of the most important matters to be considered by the purchaser of a farm before he makes a final decision. What kind of soil is it? What kind of soil parentage has it? The purchaser can usually get local advice from the county agricultural agent, but might better inquire at the U. S. Geological Survey.

Dr. C. E. Greene in *The Cancer Problem* (Greene and Son, London), says that cancer is least frequent in districts which lie on a limestone geologic formation. Professor P. Schrumpf-Pierron and Dr. Pierre Delbet issued a series of papers in Paris. These were published in the *Bulletin de l'Institut d'Egypt*, Vol. XIV (1932) and in the *Bulletin de l'Academie de Medicine*, Vol. III (1934), and they suggested a direct relationship between the incidence of cancer and the rock formation from which the soil derives. Certain rocks of a dolomitic character which are a form of limestone rich in magnesium, produce a soil which seems to be low in cancer incidence, whereas rock formations such as granites which are rich in potash have the reverse effect.

Potash is slightly radio-active and seems to be suspect as a cause of cancer. The potash-magnesium ratio in soils and plants has long puzzled scientists. If magnesium goes down then potash goes up. In many orchards due to increasing potassium, magnesium deficiency diseases occur.

These French scientists carefully plotted maps which indicated the extent of magnesium in the geologic formations of various parts of the world, and then they placed on top of it the cancer mortality figures for those regions. The agreement was startling. Egypt, a country rich in magnesium, shows an extremely low cancer rate. Certain parts of France also show a similar situation. There are doubtless other factors, known and unknown, in the cancer equation, the sulphur theory of Dr. Greene, for instance; but in this connection one must point out that sulphur is found more often in granites and the other rock formations listed as producers of inferior soils, than in limestone.

Dolomite, a limestone rich in magnesium, would, therefore, seem to be an essential soil amendment, to be used in liberal quantities in the compost heap or pit, or for top-dressing generally for all plants that do not require an acid soil.

The question may well be asked, "What shall I do if my farm happens to be located on one of the unfavorable soil formations?" Use liberal amounts of lime and dolomite and none of the strong chemical fertilizers. Make and apply as much compost as possible. I believe you can gradually build up a perfectly safe soil, the organic way.

5

SPREADING COMPOST

To return to our discussion of composts, the finished product may be applied to the fields with a manure-spreader, or forked out of a wagon as it moves very slowly over the field. Sometimes when it is time to plant a crop the compost is only half completed. It may then be necessary to spread it anyway as the process of humification will continue in the soil. It will not produce as good results as if it were composted completely, but conditions may dictate such action.

To get good results with the average crop it is advisable to use about ten tons of compost to the acre. In later years, after a general improvement in soil fertility has occurred by enrichment with humus, five tons to the acre will be sufficient. Potatoes and tomatoes do not need a rich soil, and over-use with these may stimulate the plants to run to foliage at the expense of fruit or tuber. But there is far less danger of this, with compost, than with raw manure. Truck farmers sometimes apply 50 tons of manure to the acre. The small gardener should try to make as much compost as he possibly can. A minimum for him is about half an inch over the entire surface dug in not deeper than four or five inches. He can use as much as three or four inches of the material if he can produce it.

Some farmers apply their manure or compost only when they plant corn in the rotation, figuring that it leaves a residual value sufficient for the three or four remaining years in the rotation. Experimental work done at the Ohio Agricultural Experiment Station seems to show that it is best not to use all of the manure with the corn. This work is described in the *Ohio Farmer* which states:

"L. E. Thatcher of the agronomy department of Wooster describes one set of tests in which the crop residues and manure are used in different ways in a four-year rotation of corn, soybean hay, wheat, and mixed alfalfa-clover-timothy hay. The yield records for 15 years show an average of about 2 tons of mixed hay per acre when all the manure is plowed down for corn. Where the manure was one-half plowed down for corn and one-half disked into the seedbed for wheat, the mixed hay yields were 2.2 tons per acre, and when one-half the manure was used as a winter mulch on wheat the hay yields averaged almost 2.5 tons per acre.

"Results of this experiment as well as several others, show that dividing the manure between the corn and the wheat does not reduce the yield of the corn in the rotation. One reason for this is that the better hay crop means a better sod crop to plow under for corn. In the experiment cited above, yields of corn under the three methods were practically identical at a little over 70 bushels per acre."

As soon as a compost heap is finished, it is best that it be applied to the land. The small gardener can work it in between and in the rows even while his summer crops are growing. It may be used partly as a mulch right around the plant rows. The best method is to apply compost on the surface and to cover it with a straw mulch. This will prevent the tearing out of surface roots caused by digging in the compost. Most plants have lateral roots which spread out quite far, very close to the surface. Earthworms, also, will help greatly in distributing it.

If compost is finished in the fall the gardener may dig it into the upper four or five inches of the soil. If the farmer has compost ready in the summer he can apply it where a grain crop such as wheat or oats has just been taken off, although this may be complicated by a nurse crop such as clover coming up which was planted in the grain earlier in the spring. The farmer will not find it as easy as the gardener to use up his compost quickly. He may decide to use it in the fall when planting winter wheat and barley.

If compost has to be stored it is best to move it under cover. The simplest way is to use a very heavy straw mulch or one made from withered weeds, or simply canvas. As soon as the compost is completed, certain chemical changes take place which reduce its value. Anaerobic bacteria which act without oxygen begin to act and stored heaps should be turned from time to time and watered. New machines, which will soon be available, automatically turn compost heaps. Where many heaps are made they should be covered for winter protection with straw or hay or leaves. If compost is stored in a shed or an outhouse it may be placed on a concrete floor as it is desired to slow down decomposition. The heaps can be covered with wet burlapping.

Method of Applying

It is most advisable to spread compost on a cloudy day as the sun will dry it out and some of the valuable chemicals such as ammonia which are in solution in the water will escape into the air and be lost. The farmer should plow it under or disc it in as soon as possible after spreading. Dr. Ehrenfreid Pfeiffer says, "Controlled experiments in Holland have shown that if the yield of immediately-plowed-in-manure is reckoned at 100 percent, then the yield from manure that has lain on the field three days is only 86 percent, and the yield from manure that has lain on the ground for some weeks is correspondingly less." Dr. Pfeiffer feels that (rotted) manure should "be plowed under within three hours if it is to retain its full value."

In the *Pennsylvania Farmer* of January 9, 1943, J. B. R. Dickey gives further support to the same idea. He says:

"Old-timers used to tell us that the best plan was to spread manure and plow it down immediately, before it had a chance to dry out. Science now proves they had the right idea. In a Danish experiment, run on four different crops over a term of years, if the increase in yield of the first crop due to the manure plowed down as spread were rated at 100, manure spread six hours before plowing, presumably in mild weather, gave an average increase of only 82; spread 24 hours before plowing averaged 72; and four days before plowing only 56. This is an almost unbelievable loss and it would seem very worth while for some American scientists either to verify or disprove it and to work out practical methods of reducing a national, annual fertility loss which make some others, now in the limelight, appear trifling."

If a large field is to be manured it would be advisable to do it in portions and as you place the compost in a small section, follow it up immediately by plowing. As much as half the nitrogen will be lost if the manure is allowed to lie too long.

Please note that the above experiments have been performed with manure and not compost. Manure and compost are very different things. I do not believe that the loss would be anywhere nearly as great if compost were allowed to lie instead of manure because it is not so rich in ammonia or nitrogenous compounds which might escape into the air. At any rate experimental work should be performed to test this point.

Weight of Compost

It is difficult to estimate exactly the weight of a compost heap as the materials and other conditions vary so greatly. As a very rough average, however, it may be stated that there are about two cubic yards to the ton. Weight will, of course, vary depending on the water-content of the compost. Suppose your heap is 10 feet by 20 by 3 feet high. Its total cubic footage then would be 600, before allowing for the slope of the sides. About 20% should be deducted on account of such slope which then leaves 480 cubic feet. As there are 27 cubic feet in a cubic yard we divide 480 by 27 and get 17.7 cubic yards. Since there are two cubic yards to a ton, this particular heap would amount to approximately nine tons.

Winter Conditions

The question as to when to spread compost ties in with seasonal conditions. In winter, in average medium cold to really cold climates, the biologic action in the compost heap stops, and for all practical purposes making compost in winter is not to be recommended. If one wants to have finished compost to apply in April or May it will have to be started about August or at the latest September, except in more temperate regions. Compost heaps can be assembled during warm spells in winter time, but it will be necessary to have some way of keeping a supply of earth from freezing so that it may be sprinkled in the heaps.

In green-house gardening, it may be important to have a continuous supply of compost and insulated sheds may be constructed which will permit the process of composting to proceed even during really cold spells. In such cases, very little of such a shed should be above ground. It might consist of a roof set on the ground, the entire shed itself being underground, thus preventing the heat from escaping.

6
COMPOST HEAPS FOR SMALL GARDENS

The small gardener, in making compost heaps, follows the same theory as that for the bigger heaps. Practically the only difference is in size. No compost heap should be smaller than five by five by four feet high. Any smaller size than this will result in the ratio of cooling surface to volume being too great, and in a drying of the surface layers, thus retarding the development of the bacteria and fungi.

The action of winds tends to dry out these small heaps and large heaps always show better results. In making small heaps build the second against the first and so on to conserve heat. Ordinary protection by covering with straw mulch, even to as much as one foot thick, or by canvas or burlapping is required. In using a mulch, it does not mean that you can eliminate the outside layer of heavier manure which must cover the heap to a thickness of an inch or two. The mulch or burlap goes on top.

Sometimes the gardener is in a difficult spot to supply animal manure for his heap. He can no doubt secure all the green plant matter he may need and more. Sheep and cow manure are sold in hardware stores put up in bags but are dehydrated as a rule. This is not as good as undried and not chemically treated, fresh manure. Also you may sometimes note on the bag in extremely small letters that superphosphate has been used as an absorbent. Superphosphate is one of the stronger chemical fertilizers which we do not recommend.

You can use dried blood, kitchen wastes which contain some animal matter, bone meal or fish cuttings. Dried blood may be purchased in the usual fertilizer channels. In using it instead of animal manure, apply only about ⅛th of an inch instead of the two inches recommended for manure. The exact proportion is one pound of dried blood for every 100 pounds of dry vegetable wastes.

The problem sometimes arises as to what to do when there is not sufficient material to make the heap at one time. Suppose you have grass clippings once a week. It is advisable to mix weeds and some leaves with it, the wider the variety the better. Keep adding to the pile every week. As stated previously it is best to allow green matter to wither, but not to dry, before using in the heap. If the green matter is too succulent, silage will result instead of compost. When there is sufficient green matter available for the entire heap you then proceed to make it. If, however, you decide not to wait but to construct your heap in parts be sure to cover it with a light dressing of soil —say half an inch—after you complete each stage. This should be thinned off when you begin again with the next accumulation of material. Where there is insufficient material to make a heap at one time, it might be advisable to use a New Zealand box.

Small Pit Method

Make the pit about six feet by ten or twelve and about three feet deep. If you wish, the sides may be cemented, but this isn't necessary. The floor must be of earth. Make your heap as you accumulate the manure in layers with green matter, limestone and earth, but be sure that you cover it with burlapping between each working. It is complete when it is about four feet high. Make air holes as in the regular heap.

As soon as the final height is reached turn the heap over, and again five weeks later. Pits may also be used on farms with larger size heaps and in most cases they will give better results than those above ground. From year to year, as these pits become saturated with bacteria, the action will be quicker and results more satisfactory.

Garbage Disposal

The open pit method is very effective in disposing of household kitchen wastes, but in this case it is advisable to have two adjoining pits. The size should be about three feet by six by about two feet deep. In one pit, which we shall call A, keep manure that is pretty well rotted or finished compost. If that is not available use fresh manure or peat humus. In pit B place your daily accumulation of kitchen cuttings. Then cover thinly with limestone or wood ash to help in the alkalinizing process and also to discourage cats, dogs and rats. Then give a light sprinkling of earth and finally cover with manure taken from pit A. The earth may be left out if it is too difficult to obtain, but even the lightest sprinklings will be helpful. There will be no odors and no flies.

To cover our garbage pits we use wooden frames of one-half inch square mesh wire covered with burlap. They have a handle at one end and fit loosely in the pit. They lie directly on top of the garbage or manure and work wonderfully, keeping all animals away and preventing the material from drying out by sun or strong winds. They keep the pit moist, prevent heavy rains from washing up kitchen wastes so that animals can get at them, or flies be attracted to them.

The frame should be painted with linseed oil, rather than creosote, old motor oil, or regular paints which contain materials harmful to the pit-matter. Such frames can be designed to fit individual tastes and local conditions.

When pit B is almost filled it should be turned so that the most recent garbage is placed at the bottom. The top is then given a thin dressing of manure, compost or peat humus and treated as a regular compost heap, making air holes and covering with the burlap. Five weeks later it is again turned, but this time no holes need be made. In a month it should be ready for the land. Always save a little of your finished compost for the empty pits. In time these pits will become very rich in biologic soil life and their "power of digestibility" will be great.

Handling kitchen wastes in winter presents complications. A covered pit is about the most practical method, the entire cover in this case being of wood and if possible hinged at one side for ease of handling. In this case the sides of the pit should be cemented. In order to insulate the pit a double cover at the top could be designed with an air space between. A protective mat could be placed over the cover as a further protection in very cold climates. Whatever is used, it should be made so that the cover can be quickly opened.

Some persons bury garbage in trenches and then grow plants in them. In many cases insufficient time is allowed for the thorough decay of these wastes. It is a rather dangerous thing to grow vegetables in partially decayed garbage. Some years ago in California there was an outbreak of intestinal trouble which was traced to the eating of oranges from trees which had received dressings of raw garbage. In Paris, municipal heaps are made on the outskirts of town of garbage and other organic materials. The public is permitted to take this material, but they rarely wait until complete decomposition has taken place. As a result it is said that many Parisians are afflicted with intestinal parasites. If garbage is buried in trenches, tests should be made to see that thorough decay has taken place before food crops are grown there. It may be necessary to wait at least two years before using such ground.

The danger from this partially decomposed garbage lies in the possibility that some of the decomposition organisms or some of the decomposition products may find their way onto the fruit either by dust particles or by the agency of insects such as flies, bees and wasps. One must be especially cautious with such plants as radishes and lettuce, or any other which is usually eaten raw. In the tropics the invariable rule is never to eat any vegetable that has not been cooked.

Another theory maintains that decomposition organisms or products are absorbed into the vegetables themselves through the roots—a theory that is disputed. In any case, the fact remains that it is dangerous to grow vegetables and fruits in partially decomposed garbage.

The New Zealand Box

Difficulties facing the small gardener in making compost by the Howard method can be overcome through the making of the New Zealand Compost Box. This box was devised by the Auckland Humic Club of New Zealand and works well in that country. It should prove equally valuable here. This box protects against both wind and rain.

There are many variations of this box but the simplest one is a wooden structure four feet square by three feet high with neither top nor bottom. The wooden sides consist of pieces of wood six inches wide by one inch thick. As they are nailed on, one-half inch air space is allowed between each board so that air may penetrate into the heap from all sides. The box is movable. The boarding in front slides down between two posts so that when it is desired to empty the box these slides may be pulled upward and taken out, one by one. The framework is held together by 2 x 4's.

The preferred method in filling the box is as follows: Mix the green matter, soil, lime and manure thoroughly, eliminating the layers of material. Make one air hole in the center of the box all the way down to the ground with a crowbar. If you turn the mixture twice, no air hole is needed after the second turn.

Cover the top with burlapping, or make a burlap frame with ½ inch mesh wire, for use as a cover. When the material in the box is turned over, it has to be taken out of the box in a pile and then put back again. This double operation may be saved by building a double-sized box eight by four feet square with a partition in the center. The material is forked from one into the other and a new batch is then placed into the first section. This makes it a continuous process where only one turn is given. Where two turns are used, a three-section box can be made which is twelve by four feet. It is started in section A. When A is turned into B, a new batch is made in A. When B is turned into C, then A is turned into B, and a new batch is made in A. The process is continuous. We have used a three-section box and it gives good results.

The 4 x 4 box will produce about a ton of compost. If the quantity of wastes available warrants, a larger size box may be used, 6 x 6, three feet high.

7

ANIMAL MANURES

The use of chemical fertilizers dates back to the middle of the last century. Before that time farming was carried on almost exclusively with the aid of animal manures and other residues which originated on the farm, and in many cases with composts made of decaying animal and plant matter. In many European countries chemical fertilizers were practically unknown even until very recent times. Hitler's march into Poland, for example, was followed soon after by very large shipments of chemicals, in order to increase food production. Many of the peasants had never before handled it in their lives.

In very early days man began to utilize various materials which he found on his farm or nearby, which seemed to give him better yields. The Romans used lime and when they overran England they introduced it there. England had tremendous deposits of chalk which, from then on, have consistently been used as a soil dressing.

Wood ashes also have been used as a fertilizer since long before Roman times. About 1650, saltpetre, a natural product obtained from manure, decaying organic matter and wood ashes, was discovered to have fertilizer value. Saltpetre (potassium nitrate) is used also for making gunpowder. Other materials dabbled in from time to time as fertilizers have been bones, coal ashes, charcoal, gunpowder, turpentine, tallow, oyster shells, flower of brimstone and citrated tartar, to mention a few.

In the main, however, the average farmer in olden days depended on animal manures, green-manure crops and other vegetable and animal residues. Where he followed established and well-tested cultural practice he obtained good results. Where he was lax and violated fundamental agricultural principles he was visited by crop failures. Whole countries in ancient days were turned into dust-bowls when the farmer sinned against the land.

Ferrero, one of the greatest of modern historians, has traced in *The Greatness and Decline of Rome* the decline of Roman agriculture through absentee ownership and one-crop specializing, and showed how it paralleled the economic trend of that great empire. Roman agriculture in the days when Rome was becoming great was a well-balanced, self-contained industry with sheep and oxen as the principal animals, and wheat, olives and grapes as the most important plants. The sheep furnished the wool for the clothing, the oxen were the work animals and, with the sheep, provided the meat. The lower, level fields along the rivers and on the plains produced wheat and pasturage; the rougher land grew the olive for oil or the grape for wine. The labor was largely that of the owner, or was carried out under his direct supervision. Grain was sold to the cities for articles not produced on the farm.

As the empire grew, the culture of the olive and grape became more profitable than that of grain and the growing of wheat declined. Olive and grape growing represent much hand labor and the cheapest was that of slaves and trained slaves could be bought very cheaply as a result of military conquests. The small farms were gradually given up and united as great estates where the work was done by slaves under a trained overseer who might also be only a slave, while the owner lived in luxury in the city of Rome. The country now was no longer independent. Grain had to be purchased abroad and the great capital was dependent on the arrival of grain ships from North Africa.

Ferrero used to visit America frequently. He said that he could best study in this country the subject nearest to his heart, the decline of a great republic. And agriculture was one of the most important factors in that decline.

The greatest virtue of the old-fashioned farms was that they were self-contained and had a balanced complement of both crops and livestock. Monoculture, the running of a one-crop farm with practically no animals, was rarely seen in America—except in the South where the worst cases of wornout land are to be found today. It took scientific agriculture with its test tubes, chemicals and laboratories to encourage farmers generally to run a farm as a factory, in many cases on a 100% chemical basis. What happens in such cases is land "mining."

The soil soon loses its structure. Plant diseases break out. The crops become deficient in nourishing qualities and other troubles are encountered.

Luther Burbank, whose name is a symbol to horticulturists, built a basic foundation for his experimental work by loading his soil with a tremendous amount of animal manure. On his four famous acres at Santa Rosa, California, he spread eighteen hundred loads of barn manure before he started. This is an enormous amount of organic material for only four acres and demonstrates his leanings towards the old-fashioned way of preparing the soil as contrasted with his ideas in connection with producing new kinds of plants.

Burbank had been raised in New England where the soil is normally poor and from his early experience he had learned that in order to find out what a plant really can do, it must be given a full and rich diet and the best thing he knew for this purpose was the material with which the poor soil of his native state had been enriched—i.e., with animal manure.

More than one billion tons of animal manure are produced in this country every year. It is estimated that this amount contains plant-food values that are more than six times that of the artificial fertilizers used in the same period. By following the Howard method of composting, this billion tons of manure can be increased 300%, and, at the same time, food values for growing plants can be improved. It also means that a given number of cattle can maintain the fertility of more than double the number of acres under methods commonly used now. This does not mean that we need fewer animals. The more we have, the greater use can be made of all their products.

Animal manure by itself is an unbalanced fertilizer—unbalanced chemically as well as biologically. There may be too much urine and too little cellulose, or vice versa, thus preventing the microbes from performing their work efficiently. By following the Howard method of making compost, not only will the proper biological processes proceed smoothly in the decomposition, but the materials—plant and animal—of the compost heap, provide every needed soil-food element.

Animal manure when used alone decomposes in the soil much more slowly than plant matter, although certain types of plant matter, such as pine needles, wood shavings and sawdust, are slow to decay. However, when manure is composted along with plant matter by the Howard process, decomposition takes place much faster. The association and position of the various materials, and the other factors, such as aeration and heating, bring about a more accelerated break-down in the animal matter than would occur if it were plowed under.

When manure is not composted, but is permitted to weather under all kinds of unfavorable conditions, the "bandits"—the anaerobic bacteria that can live and work without oxygen—destroy much of its valuable nitrogen, which goes up in thin air. One method of saving some of this nitrogen is to use a large amount of bedding for cattle. Conversely, where little bedding is used and manure is heaped in piles, "fire-fanging" occurs—a form of very destructive fermentation. Such manure has very little value. But the chief reason why sufficient bedding must be used is to absorb the urine. Fifty per cent of the value of manure is contained in the urine.

Ordinary manure-heaps smell, attract flies and sometimes become so saturated with rain during heavy storms that much of the valuable nutrients are washed away, sometimes as much as one-half. Manure heaps are often carelessly piled where surface drainage water washes into them. Some farmers shovel manure through a window of the barn and let it accumulate under the eaves where excessive rain water washes valuable elements away.

Bulletin 92, entitled *Farm Manures*, issued September, 1932 by Clemson Agricultural College of South Carolina, states:

"With the present lack of attention given to saving manure, the loss between animal and field is enormous. It is doubtful that 20 per cent of the manure produced in South Carolina is actually placed in the fields."

Farmers often carelessly destroy manure and then spend hard-earned money in purchasing artificial fertilizers, because it seems like less work. In the long run, they are making more work for themselves and their families. I know of cases where farmers did not avail themselves of opportunities to get free manure because it involved trucking a few miles. Many farmers actually sell manure. In most cases it will be found that they are tenant farmers. The land is not theirs. They do not have the incentive to invest in fertility, for next year they may have to move to another patch. Recently I stopped to check at two farms that had a sign out "Manure for Sale." Both of them were run by tenant farmers. Sometimes a very large dairy will sell manure because they do not grow much of their feed. Their cows are largely fed on concentrates. In China manure is valued so highly that every bit of it is scraped off the roads and used.

Manure should never be used raw. The recommendation of many Government authorities to haul daily accumulations from dairy barns direct to fields is greatly to be deplored. A method recommended by some dairy authorities seems to have excellent possibilities. There should be a separate milking barn where the cows are brought for milking. They are then turned into a covered shed which is kept well-bedded. Liberal application of bedding which is trampled down by the cows will preserve the urine and will prevent destructive fermentation. Another valuable method for conserving manure and permitting it to rot before using is the pit system described in the section, "Short Cuts in Compost Making." But to preserve most of the values in the manure, it should be composted with plant matter as soon as possible.

There are many reasons why raw manure except in the rarest of emergencies should never be applied as a fertilizer unless it is done seven or eight months before the crop is grown and given a chance to decompose. The *first* reason is the same as that which applies to the plowing under of a raw green-manure crop. The bacteria and fungi, in having to break down so much raw matter, must consume a great deal of the available nitrogen and other food elements in the soil to furnish energy, thus depleting it for use by the new crop. In many such cases very disappointing results are obtained and benefits are not observed until the year following, when the enriched dead bodies of these microbes become available.

Second: Weed seeds in the manure are not killed off, as in the compost heap. Thus many farmers by applying raw manure actually plant a crop of weeds as well as insect eggs and spores. Where fresh manure is obtained from outside sources new species of weeds may be introduced to the farm as well as certain disease organisms.

Third: Where a herd of cattle has Bangs or other diseases the organisms are planted in the soil. The high heat of the composting process kills off these dangerous organisms. The germ of Bangs disease causes undulant fever in man. Its organisms can live in the soil for many years. *U. S. Farmers' Bulletin No. 1568* recommends that all rabbit manure be burned to prevent re-infection. This is a tremendous waste; fortunately other Government bulletins recommend its use as a fertilizer. Careful composting of such manure would make it absolutely safe as a fertilizer. It is dangerous to use it raw.

Fourth: Plant and human health can suffer because of the absorption by the plant of half-rotted albumin and other dangerous substances from raw manure. Clubroot is a common plant disease aggravated by the use of raw manure.

Urine is an extremely important fertilizing substance, but it, too, should not be used raw. Where it is applied direct to the land it can have a caustic effect. Four-fifths of the potash in the manure is in the urine. It contains half the nitrogen but practically no phosphorus. Urine is rich in the most important substances needed to create a fertile soil. Many farmers shamefully waste animal urine, not realizing the value it has if properly employed.

There are growth-promoting substances in manure, called *auximones,* which are roughly comparable to vitamins and hormones in their effects. Very little is known about them, except that they are abundant in the growing tips of plants and it has been demonstrated that manure from animals feeding on fertile land is richer in them than that from animals grazing on poor land. The possibility exists, however, that these auximones are in reality none other than varieties of vitamins.

In the *Journal of Medical Research,* Volume 14 (1926), there is an account of an experiment at Madras, India, related by Sir Robert McCarrison, M.D., which proved that where animal manure was the fertilizer medium the crop of grains contained more vitamins than where commercial fertilizers were used. There are many other experiments which prove the same thing and authorities attribute these results unquestionably to growth factors in animal manure.

Professor W. H. Schopfer, Director of the Botanical Institute, University of Berne, Switzerland, one of the world's greatest authorities on the subject of plants and vitamins, says in his book, *Plants and Vitamins:*

"the action of animal excreta cannot be overlooked. Because of their high vitamin content the feces of horses and other large animals serve as excellent media for the culture of numerous fungi . . . finally, manure applied to the soil represents another source of growth factors. Bonner and Greene (1938) found the following figures for a few manures: Arizona steer manure 0.13 mg., local steer manure 0.08 mg., dairy manure 0.13 mg., of thiamin per kilogram. The presence of growth factors in natural manures must contribute to their superiority as compared with artificial fertilizers. This superiority has been established for a long time."

As a farm continues to be operated in the organic manner and its soil grows more and more fertile, its manures likewise grow richer. Eventually much less manure is needed. Where ten tons of compost per acre may be needed at the beginning of a program, in five or six years of operating with compost only five tons per acre may be sufficient.

It stands to reason that manure from a sick animal is not up to the quality of that from a healthy one. In the same way manures from cattle that are fed high quality feed are much superior to that of cattle depending on an inferior food ration. It is this difference in quality of manure that effects in direct ratio the quality of food and its taste, when grown in soil with such manure. For the very same reasons weeds, leaves and other green matter originating on an organically-operated farm, are superior in soil nutriments to that from a farm where chemicals are relied upon.

The best manure, from the point of view of ease of fermentation or decomposition, comes from feeding cattle hay, grass, alfalfa, plus a reasonable amount of grain and a small quantity of a supplement like soy-bean meal. Where too much concentrate is fed, cottonseed and soy-bean meals, etc., the manure will be too sticky and more resistant to decay.

The analyses of various kinds of manure such as horse, cow, steer, pig, sheep, chicken, and others, show a great deal of variation as to nitrogen, phosphorus and potash content. Chicken manure is richest in all three, for fowls do not excrete urine, and therefore might serve its purpose best by being mixed with other manures. Its habit of burning plants when used fresh shows the necessity for composting it before use. A method of effecting a natural and gradual composting right in the chicken house is described in Part V, the section, *A New Type of Chicken-Droppings Pit.*

Many poultrymen sell their manure and do not produce their own chicken feeds. With the aid of this rich organic material they could produce a far superior kind of feeding matter than they can purchase. They would thus have much better results with their chickens. The nutritive quality of the eggs and meat would be higher. I have even seen cases where poultrymen had difficulty in disposing of their accumulated chicken manure, having to accept a very low price for it.

Horse manure has the characteristic of easily "fire-fanging" if allowed to accumulate in heaps. To counteract this tendency it should be watered liberally. Horse manure has the good quality of preventing the harmful actions of denitrogenizing bacteria. That is why good results will be obtained if it is mixed with other manures, even when making compost heaps. Pig manure is highly concentrated and is best when mixed with the excretions of other animals.

There are a few cautions in connection with the use of animal manures if you insist upon applying them to the land without composting. Fresh manure is always acid. For the average vegetable and farm crop a too acid condition may interfere with plant growth. Where too much raw manure is used for tomatoes it will result in maximum stalk and plant growth and minimum fruit. This does not occur when well-finished compost is used. Be careful in purchasing horse manure that it isn't "spent" manure.

I have noticed for sale in some of the stores bags of cow manure, with a notation in very tiny letters near the bottom that superphosphate has been used as an absorbent. This is no doubt due to the fact that it is standard practice in dairy-barns to sprinkle superphosphate in the manure every day as a deodorizer and to lock in the nitrogen. I do not recommend it.

To sum up; if possible, never use raw animal manure. If you cannot compost it let it rot under conditions that will preserve most of its nutrients. But for superior crops make compost of it. Balfour in *The Living Soil* describes an interesting experiment: "A large scale test, involving forty fields, was made on the late Sir Bernard Greenwell's estate of humus manufactured by the Indore process versus best quality farmyard manure, that is to say, well-rotted manure that was practically pure dung. Load for load the compost showed in every case better results." This experiment was described in detail in the *Journal of the Farmers' Club of London* in 1939. I would rather use five tons of well-made compost than fifteen tons of raw manure.

I am not discussing the use of human excrement as a fertilizer in this book for two reasons. First, at this stage in our agriculture there is no need for it in this country. In China and other parts of the Orient it is extensively used, having been made necessary by centuries of intensive farming and overpopulation. Our problem in the United States is, first to find ways to use the millions of tons of vegetable and animal wastes which today are thrown away, burned up or otherwise destroyed. Secondly, the subject is highly controversial. Many medical authorities feel that by employing human excrement, disease organisms such as typhoid and hookworm continue to live in the soil and may become attached to the plant and be eaten with it—as on lettuce leaves. Sir Albert Howard, however, states that careful composting will remove all dangers. Cases that one hears of, where its use caused outbreaks of disease, may have been due to its use on the land without benefit of composting. Careful research is needed in this field, before any conclusions may be drawn. Balfour and Howard, in their books, cite careful composting programs and techniques for the extensive use of this "night soil" and other town wastes. I refer the reader to them for a thorough discussion of the matter. Both are extraordinarily careful scientists and practical farmers, and they come to no hasty conclusions or generalizations.

A spectacular instance of the use of manure and other organic matter in the elimination of the root-rot of cotton is graphically described by J. Sidney Cates in the *Country Gentleman* of February, 1944. My quotations are from different sections of his article, but are not out of context:

"Little has been said about this new cure for root rot. Perhaps this is because the remedy, in its outward respects, is so very simple.

"The cure for root rot of cotton consists in nothing more than turning under leguminous organic matter in the fall or spring before the cotton crop is planted. A heavy dose of stable manure will affect the same end, though the latter is seldom available in sufficient volume to be a large factor under most root rot infested cotton-farm conditions. But this leguminous-organic-matter treatment of cotton soils has long been known to be highly profitable, even when no root rot is present.

"It is not through the chemical action of this rotting material that the cure is brought about. The answer lies in the fact that teeming bacteria in the soil, these bacteria no doubt embracing thousands of strains, take on renewed life under the stimulation of this, their seemingly pet food supply. Some of these stimulated bacteria strains are dread enemies of the root rot fungus. The bacteria not only destroy the fungus on contact, presumably as certain bacteria destroy our own body tissue, but the antipathy between these two organisms goes much further. In artificial-media cultures the root rot fungus dies even on the approach of the enemy bacteria growth.

"The discovery leading to its control comes from C. J. King, Superintendent of the U. S. Department of Agriculture field station at Sacaton, Arizona. King observed one summer in a section where alfalfa had been grown, and in another section where stable manure had been added, that there was little or no root rot damage while alongside, to the very sharp margin of where these treaments had been given, the cotton crop was almost devastated by root rot.

"But this new and successful root rot control method—that of altering the soil microbiology by feeding it with decaying leguminous material—is far bigger than the mere announcing of the abatement of a curse to Southwest cotton, devastating as the curse has been. *Fertility* is a chemist-proffered term. The word 'productivity' gives the realistic picture to the practical farmer. For years we have been twiddling over the proper proportion of NP and K and more recently over so-called trace elements, while the understanding of the broad phases of the soil-productivity mystery is not only confused, and unknown, but grossly neglected."

8

SHORT CUTS IN COMPOST MAKING

The making of compost heaps has been modernized recently by a device called a manure loader which attaches to a tractor. This machine, which is not too expensive takes the back-breaking labor out of making and turning compost heaps. It makes organic farming practical for the twenty acre or thousand acre farm. Where it takes a half day to turn over a large heap by hand, the machine will do it in a few minutes.

These manure loaders may be used to assemble and place the materials in the heap, turn it over when necessary, and also, finally, to load it onto the manure spreader. One company, in advertising its machine, states that it will also stack hay, pull fence posts, build dams, shovel snow, lift rocks, haul hay, raise small buildings and do a half dozen other jobs around the farm. The cost of these machines ranges from about $150. to $350., and thus are within the means of the average farmer. This should be wonderful news to farmers who have hesitated to adopt the organic method on account of the labor involved. Names and addresses of concerns manufacturing these machines are given at the end of this chapter. Several of these companies may be able to deliver these machines right now. Others will not be in a position to do so until after the war.

Until every farmer who wants to can get one of these machines, there will be many who will prefer to follow the organic method of farming without turning over the heaps. There are short cuts and compromises which many can put into effect. Some may be used to good advantage even by every one.

Sometime ago I described the organic method to a neighboring farmer. When I met him a few weeks later he said he and his wife had thought about it a great deal afterwards and that I must be right, according to their own experience:

"In the old days, when there weren't any State sanitary regulations affecting the running of a dairy farm, we had a big shallow pit right outside the barn, and the manure was put there every day. It was allowed to accumulate and rot for a long period until planting time when it was applied to the land in a fine, darkened and rotted condition. Today the law prevents that so we have to take the manure daily and apply it right away to some field. Now, my wife and I have often spoken about how much finer our crops used to be in the old days but we never could think of the reason for it. After what you tell me, we are sure of the cause."

In Thomas J. Barrett's book on the earthworm there is a chapter that describes a natural method of making compost used by Dr. George Oliver's grandfather, part of which is given here because it is so apt:

"In the center of the barnyard was the compost pit, which, in the light of modern technical knowledge, I now know to have been the most perfect and scientific fertilizer production unit I have ever known. This pit was 50 feet wide and 100 feet long and had been excavated to a depth of two feet. At each end, evenly spaced from side to side and about 20 feet from the end, was deeply anchored a heavy log post, probably 12 to 15 feet high, with an overhead cable anchored to each post and running to the barn. On these cables were large traveling baskets. Thus, each morning the manure from the great barn was transported in these baskets to the compost pit and evenly spread in a uniform layer, by means of the post in each end of the pit, the manure being dumped at the most convenient point for proper handling. By means of the automatic dump baskets and overhead trolley, it was possible to quickly clear the barn of the night droppings and deposit the material in the compost pit without any loss of the valuable elements of the fresh manure.

"To one side of the barnyard and at a higher level than the floor of the yard was located the ice pond. This pond was so arranged that it could be filled from a flume leading by gravity from the creek to one end, while at the lower end a spillway was provided so that the pond could be drained. At the proper season, the ice pond would be filled to the correct level and when the ice formed to the proper thickness the annual harvest of ice was cut and stored in the ice house, to provide an abundance of ice for all purposes the year round. The bottom of this pond was formed of a fine-textured red clay. Each spring the pond was drained and with teams and scrapers many tons of this clay was scraped out and diked around the borders of the pond, to weather for use on the compost heap.

"And now enters the earthworm. For more than sixty years my grandfather had farmed this 160 acres without a single crop failure. He was known far and wide for the unexcelled excellence of his corn and other grain and a large part of his surplus grain was disposed of at top or fancy prices for seed purposes. As previously mentioned, the pit was 50 by 100 feet, excavated to a depth of two feet especially to provide a great breeding bed for earthworms. Literally millions of earthworms inhabited this pit. Part of the daily routine of the farm work was the building of the compost heap. Each morning the barn was cleared of the droppings of the previous 24 hours, which were transported in the dump baskets on the overhead trolley and dumped on the compost heap, where it was evenly spread over the surface.

"The flock of chickens everlastingly scratched and worked in the barnyard, assisted by the ducks, gleaning every bit of undigested grain that found its way into the manure and, incidentally adding about twenty tons of droppings per year to the material that eventually found its way into the compost heap. The cattle and sheep used to wander around the four straw stacks and bedded under the shelter of the stacks, adding their droppings to the surface and treading it into the bedding material. From time to time, the entire barnyard was raked and scraped, the combined manure and litter being barrowed to the compost heap and distributed in an even layer on the surface. As the compost was built up to a depth of twelve to fourteen inches, several tons of red clay which had been scraped from the ice pond were hauled in and distributed in an even layer over the entire surface of the compost. Thus, the variety of animal manures from horses, cattle, sheep, hogs and fowl, mixed with the bedding material of wheat straw and oat straw, alternated in the compost heap in layers of twelve to fourteen inches thick, with layers of red, fine-textured clay, rich in parent mineral substance.

"Meantime, beneath the surface the earthworms multiplied in untold millions, gorging ceaselessly upon the manures and decomposing vegetable material, as well as combining with this diet the mineral clay soil, and depositing their excreta in the form of castings—a completely broken down, deodorized soil, rich in all the elements for plant life. From time to time, as careful inspection by my grandfather would determine, the compost would be watered from the flume leading from the creek, thus providing the necessary moisture to enable the earthworms to work to the greatest advantage in their life-work, that of converting compost to humus. By the time the spring thaw arrived, the season of the annual plowing, the earthworms had completed their work in the lower layers of the compost heap. The top layer of the heap would be stripped back, revealing the perfect work of the worms. What had originally been a damp, ill-smelling mixture off manure, urine, and litter, was now a rich, dark crumbly vegetable and animal mould earth, with the odor of fresh-turned earth.

"This material was not handled with forks, but with shovels for there were no dense, burned cakes of half-decomposed manure. My grandfather would take a handful of the material and smell of it, before pronouncing it ready for the fields. No material that was not completely broken down was allowed to go into the fields and the 'smell test' was a sure way of judging material. When perfect transformation had taken place, all odor of manure had disappeared and the material had the clean smell of new earth. At this time of year, at the beginning of the spring plowing, the compost heap was almost a solid mass of earthworms and every shoveful of material would contain scores of them and as I now know from years of experiment and study, every foot of this material contained hundreds and hundreds of earthworm capsules, which would within a few days of burial in the fields hatch out from four to twenty earthworms each, to immediately begin their life-work of digesting the organic material of the soil, mixing it with much earth in the process, and depositing it in and on the surface of the earth in the form of castings—a finely divided, homogenized soil, rich in the stored and available elements of plant food in water-soluble form.

"In the annual distribution of the fertilizer, my grandfather never completely stripped the compost pit. One year he would begin the hauling at one end of the pit, stripping back the top layers of material which had not been broken down, and leaving a generous portion of the other end of the pit undisturbed as a breeding and culture ground for the remaining earthworms. After the hauling of the fertilizer was completed, the surplus top material was evenly distributed over the entire pit for the mother substance and the new compost heap was thus begun. The following year, the hauling would begin at the other end of the pit. Thus, by alternating, there was always a large undisturbed breeding ground to conserve the earthworms for their important work."

Fully within state regulations, any farmer can construct such a pit at a proper distance from the barn. Or he can bear in mind the particular field where he intends to apply the manure and can haul it there, making heaps similar in shape to compost heaps and covering them with some earth. Some farmers cart their manure daily to the fields but make the piles which are left over the winter too small and much of the nutrients are lost. The combined action of sun, wind, rain and frost on these little heaps causes serious losses. The piles should be large and have a covering of earth which will help to protect them against the elements. They should be placed on a high spot which will not be washed by running rain-water. Barns are usually so constructed that the urine of the cows is drained into a pit. The farmer can greatly enrich his manure heaps by impregnating them with this liquid.

The farmer can go a step further now if he wishes. He can make regulation compost heaps in the field; that is, in addition to manure he may add green matter such as weeds, leaves, grass clippings, etc. He should then add some earth and lime or wood ashes and follow the instructions in regard to the method of assembling these materials. But he doesn't have to turn the heaps. Since he doesn't have the time to do that he will have to wait longer for the heap to decay. To do a good job without turning the heap requires a year or two, but the average farmer prefers to wait no longer than six months before applying it on his land. Even at this stage it will be much better than raw manure. By applying green matter to the heap, he is multiplying his available fertilizer by two or three hundred percent. He is also aerating the heap as well as adding nutrients, for animal manure by itself, as we have seen, is unbalanced and is not a complete fertilizer. When ready to use the heap, he peels off the outer layer and uses it to start a new pile.

Another short-cut, if necessary, would be the placing in the barn of extra heavy bedding for such cattle as steers. Ordinarily, manure contains straw or other bedding to which is added other green matter in composting. If you are not going to make compost, you can let your cattle do the work for you, by giving them extra-heavy bedding and having them thoroughly mix the mass together by treading. You will then have most of the ingredients of the compost heap. In this case do not clean out the barn as often as you do ordinarily. When removing the manure, place it in neat piles and allow it to rot. We can go still further by sprinkling earth mixed with lime or wood ashes in the bedding every day or so right in the barn. We then have all the ingredients of the compost heap, animal matter, green matter, earth and a neutralizing base.

In a later chapter describing a new type of chicken-dropping pit there is mentioned a system of leaving the manure in the steer-barn all winter. New straw is put down about twice a week and the steers trample it in thoroughly. As time goes on the manure rots from the bottom up, due to bacterial action and in time becomes fairly well rotted. We have followed this method for the last two years and have not cleaned the barn out for over eight months at a time. The accumulation of bacteria in the bottom layer aids in quickly breaking down the manure so there is no unpleasant smell.

I should like to quote here from an article in a recent issue of *Organic Gardening*, entitled "Yorkshire Foldyards," by F. C. King:

"The system you describe of keeping steers in a barn has been an accepted principle there (Yorkshire, England) for centuries, the only difference is that where you speak of a barn we, in our part of the world, speak of covered-in fold-yards. Another difference may well be that, whereas you only mention steers, we run pigs, calves, lambing ewes, etc., in these fold-yards. Poultry also have access so that we get a very good balanced mixture of dung and urine.

"Generally speaking our fold-yards are empty round about April—all the dung by that time being cleared out and on the arable land. Usually the fold-yard is in the form of a square or rectangle which is completely roofed in, and all stables and cow-sheds, pig sties, etc. are built on the outside, but always with one door into the fold-yard, so that when these places are cleaned out—which in the case of the milking-cow sheds is every day, all the dung and soiled bedding is taken through the door into the fold-yard. During the summer months, when most of our stock is out, little manure is made, the bulk being produced in Winter when practically the whole of the livestock belonging to the farm is in-doors. The centre of the fold-yard is always much lower than the sides, forming a pit.

"The side walks, which may be anything from nine to twelve feet wide are cobble-paved, which I can best describe as a lot of small hard stones set close together on edge, exactly like a lot of our small market town streets and market places which some of your readers may have seen. The base of the fold-yard, however, is seldom paved but is just the bare earth. You will at once see the benefits to be derived from this system. All through the winter the level of the centre gradually rises and is a mixture of all animal excretions and urine, together with a certain amount of soiled bedding. Natural composting is carried out on the floor of the yard, which is abundantly supplied with soil-bacteria and microbes. A steady even fermentation is excited all the time with an absence of offensive odours.

"Large troughs are scattered about for the feeding of the various stock. Steers—or as we call them bullocks—are fattened for Christmas sales, also many pigs which run about amongst the stock, picking up a lot of food which the stock drop, and generally acting as scavengers. In our country a Yorkshire farmer is considered a very good man indeed, and I hold the belief that a lot of his success is due to the care and the general management of his manure. In other parts of England this is not so, and the spectacle of manure being heaped up outside cow-sheds, etc., is all too common and is to be deplored. Under these conditions much of the original goodness of the manure is lost and more particularly the urine, which is far richer than the solid dung. These heaps develop a most unpleasant smell, and are a breeding place for all kinds of flies. Heavy rains wash out a lot of the goodness which finds its way to the drains by devious routes. Wind and sun alike interfere with the fermentation. The ammonia evaporates and the final product is unsatisfactory."

There is a method called *sheet composting* by which compost is made in the fields in one application. Sir Albert Howard describes this in the *Agricultural Testament,* as follows:

"In exceptional cases . . . it is possible to carry on the manufacture of humus in the soil without any risk of temporary failure. One British example may be quoted. On some of the large farms in the Holland Division of Lincolnshire peas are grown as a rotation crop with potatoes. The problem is to manufacture humus before the next crop of potatoes is planted. This has been solved. Early in July the peas are cut and carried to the shelling machines where the green seeds are separated and large quantities of crushed haulm (vines) are left. Immediately after the removal of the peas the land is sown with beans. The crushed pea haulm is then scattered on the surface of the newly-sown land followed by a light dressing of farm-yard manure—about 6 or 7 tons to the acre. The beans grow through the fermenting layer on the surface of the soil and help to keep it moist. While the beans are growing, humus is being manufactured in a thin sheet all over the field. At the end of September, when the beans are in flower, this sheet composting on the ground is complete. The green crop is then lightly ploughed in together with the layer of freshly prepared compost. Humus manufacture is then continued in the soil. The beans under these conditions decay quickly; the process of humus manufacture is completed before the planting of the next potato crop."

Sheet composting may be resorted to in connection with a hay crop about to be plowed up for a cultivated crop. During the last year in hay, directly after the first cutting, a layer of manure is applied to the field, and the grass grows right through it.

The degree of success in sheet composting depends to a great extent on the fertility of the soil, and it will do poorly in an extremely dry season.

Sir Albert Howard in a communication to *Organic Gardening* tells how to save time making compost if one cannot follow the original method in every detail. He says:

"This can be done by greater care in the original mixture. If great care is taken in assembling a heap, which, we will say, finally settles to a height of about 4 feet, turning could be avoided if the upper 2 feet were used first and the lower portion used after an interval, because my experience always has been that it is the portion of the heap nearest the atmosphere which ripens first, because air will percolate into a fermenting heap to a depth of about 18 to 24 inches quite easily. The result is the upper portion is ready, but the lower portion is not, so that if the upper portion is shovelled off first and the lower portion is left for another month, we might get very useful results in this way.

"Labor can be saved in compost-making by a careful assembly and mixing of the materials. In this connection you will be interested to know that as the soil becomes fertile its digestive power as regards organic wastes laid on the surface increases very rapidly; a time comes when a fertile soil will eat anything when spread on the surface.

"What I meant about mixing the materials as they go into the heap is this: instead of sandwiching them and keeping the soil, vegetable wastes and animal wastes separated, to mix them as one goes on, so to speak. It can easily be done in this way. If we start with a six-inch layer of vegetable wastes, put on a two-inch layer of farm yard manure, and a sprinkling of earth, and then fork up the layers, they get mixed quite a lot and fermentation is very much better and more complete than if the material is left as separate layers. By doing this we need only turn once. In fact, when the New Zealand Box is used, if we mix the material in the box, as I have suggested, all that is needed is to turn out this material into another empty box by the side and let the fermentation complete itself in the second box; then no more manipulation is necessary."

Manure Loaders

Manure Loader, John Deere Company, East Moline, Ill.

Hydraulic Farmhand, Superior Separator Company, 1179 15th Avenue, S.E., Minneapolis, Minn.

Horn-Draulic Loader, Horn Manufacturing Company, Fort Dodge, Iowa.

Combination Sweep, Loader and Stacker, Gunning, Inc., Frankfort, Ind.

The Haymaster, New Method Equipment Company, Indianola, Iowa.

9

ADVANTAGES OF COMPOST FARMING

To summarize, I am going to list thirty-six reasons why compost farming is superior to farming or gardening with artificial fertilizers. When anyone says that making compost heaps is too much work, he overlooks the fact that the work is not *extra;* that is, to be added to the sum total of the work on the farm. It is work that is taken from other operations on the farm, and with the lapse of a few years the total work on the organically operated farm will be much less than before.

1. *The general fertility level of the farm or garden is greatly improved by the organic method.*

2. *This method improves the soil's mechanical structure* which includes its granulation, tilth and increase of pore spaces. It is a great improver of heavy clay soils. Sandy soils are made more cohesive, retain moisture better and do not dry out so rapidly.

3. *It makes for ease of cultivation.* Individual cases show that where it used to take nine hours to hand-hoe an acre of potatoes, it now takes only seven. Another case shows that where it used to take four days to hoe beet fields, with the softer, looser organically treated soil it now takes only 2.9 days (Dr. Pfeiffer). On a 437 acre farm, operated by this method without any chemical fertilizers, a farmer was able to dispense with one team of horses, thus saving feed, stable space, grooming labor, etc. Does this not compensate for the labor of making compost heaps?

4. *It eliminates valuable waiting time.* The farmer can get back on the soil quicker after a rain. Because of the spongy, porous soil structure there is less mud and stickiness, or soil broken into clods. The soil is more mellow. Gardens that were formerly too sticky to work when wet, and which immediately baked hard as a brick, and formed large cracks in the surface when dried out, became easy to work after being handled in the organic manner. Hand cultivators push through such soil much more easily. Where land has been treated with artificials, and gets the benefit of only occasional, small amounts of organic matter it tends to become harder and harder.

5. *It increases the soil's water-holding capacity.* Estimates go all the way from a 20% to 50% increase in the soil's ability to absorb and hold rain water.

6. *It prevents soil erosion and reduces flood hazards.* I have seen the benefits on my own farm where the water, in heavy rains, does not come dashing along with any where near the volume or violence that it used to. A great deal of money could be saved in engineering devices such as the building of drains, contours, dams, and terraces if farmers were to improve the structure of their soil by using more compost on their land. According to Howard, "this is the real way of stopping soil erosion."

7. *It prevents hardening of the surface soil by driving rains.* Where

soil is hard and has poor mechanical structure, the impact of a hard rain will create a surface crust. On some farms you can actually pick up large hard crusts from the soil surface. The porous, spongy structure of an organic soil prevents this.

8. *The earthworm multiplies greatly,* because organic matter is its natural food. Artificials kill or drive them away. The earthworm is one of the best friends of the farmer, aerating his soil and actually manufacturing top-soil. Earthworms have been known to add 1/5th of an inch of top-soil to the land in only one year.

9. *It multiplies the microbial population of the soil.* Bacteria and fungi increase tremendously when soil is enriched organically, aiding the growing processes enormously. Better aeration and more constant moisture greatly encourage their increase.

10. *Land can safely be plowed more deeply.* Where land is organically rich and has the proper mechanical structure, you can do almost anything with it. The top-soil layer becomes thicker and you can plow deeper.

11. *Hard-pans will not form.* If there is a hard-pan it will gradually disappear, especially with the aid of earthworms who will gradually crumble it up. When we first purchased our farm there was one depression where rain-water accumulated and didn't seep into the ground for days because of such a hard pan. Now this never occurs at that spot.

12. *There is no danger of a plow-sole.* The same tendency of chemicalized soils which make them bake hard and form clods creates a plow-sole at the point where the plow shears the soil. This is a hard pan-like formation. This does not occur in an organic soil because of its softness and mechanical structure.

13. *Heavy machinery does not compact the soil as much.* An organic soil has a springiness which makes it rebound after the weight is removed. On soils that are hard-packed the weight of tractors, trucks and heavy combines only makes this condition worse.

14. *The soil has much better aeration.* This is quite obvious and is one of the most important requirements for optimum soil fertility. Aeration is made possible by the proper crumb structure of the soil as well as by the burrows of earthworms. The pore spaces permit a greater surface for the film of soil water to adhere to. In well-aerated soils ground air is kept in motion by the wind blowing on the surface. Such air movement helps check pathogenic (dangerous) bacteria. Good aeration is necessary for the proper development of roots. Some soils suffer from asphyxiation and the roots practically strangle. This has been proven to be a specific forerunner of plant disease.

15. *Soil made darker by humus absorbs heat more quickly and more effectively.* This is the principle of color in insulation. White keeps heat away and the darker the color becomes the more heat it absorbs. This enables the land to warm up much earlier in the spring so that work can start and seeds begin to germinate. The bacteria start working as soon as the temperature goes up. The amount of moisture and air space in the soil also has an important bearing on soil temperature.

16. *Dry weather advantages.* Ordinarily, just as one doesn't plow when the soil is too wet, one cannot plow when it is too dry, or the soil will blow away. The organic system which stores up moisture doesn't have this disadvantage. In dry weather the land cools sooner in an organic soil, and more dew is precipitated. Under drought conditions an organic soil will fare better due to its stored up water.

17. *It may actually improve rain conditions.* Robert H. Elliot, who followed the organic system, with the addition of a method of growing deep-rooted crops such as chicory and orchard grass, says in *The Clifton Park System of Farming:* "It is important to note too that the air passing over a humus-fed soil, would be cooler and moister than air passing over a mineralized soil. The dew-fall therefore would be greater and when the land throughout the country becomes generally humus-fed, the rainfall would be more advantageously distributed, and fall over a greater number of days in small showers instead of in heavy falls as in the case when land is clothed with forest."

18. *It transpires less water through the leaves.* In other words not only does it store more water, but it gives up less of it than a hard-packed soil. As a rule it requires several hundred pounds of water to be lost or transpired through the leaves of a plant for every pound of dry matter produced. In *Research Bulletin No. 6* (1915) of the Nebraska Agricultural Experiment Station, it is shown that on good soil, well-manured, the loss of water is only about a half of what it is on an unmanured soil. This is a point of vast significance, especially in regions where the rainfall is deficient at some time during the growing season, and this condition applies to the major part of the great Mississippi Basin.

19. *The manure produced from cattle fed on an organically operated soil improves in quality from year to year.* It becomes progressively richer so that eventually less is required. In about fifteen years a point is usually reached where the farm or garden can get along without any fertilizer help whatever for a few years. Manure from a chemicalized farm has far less value.

20. *Making compost by the Howard process increases the available manure by 300 per cent.* This is obvious, since ordinarily only the

available manure would be used, whereas by our process six inches of grass, leaves, weeds, etc., must be used for every two inches of stable manure.

21. *Compost heaps preserve all the food elements in the manure.* Due to the manner in which the green matter, manure, earth and ground limestone are interlayered in the compost heap, there is little loss whereas in the old-fashioned manure heap, sometimes more than half the values are leached out. There is a good deal of nitrogen fixation during the ripening process in a properly made compost heap. The Howard type compost heap is made in a manner that takes advantage of the natural action of soil bacteria, fungi and moulds. There are also no smells or flies as found in manure piles.

22. *Composts have a residual effect.* In experiments with ordinary manure 48% was used up the first year, 24% the second, 15% in the third and 13% in the fourth year. But in using composts some residual effects were felt as much as 15 years later. This builds up what farmers call "high condition."

23. *When following the organic system your grounds look neater.* Weeds are cut down periodically for use in the making of compost. When cut this way they do not go to seed and spread. Lawn cuttings, trimmings and other matter usually left around is carefully gathered.

24. *Weeds can be cultivated more easily out of an organic soil.* In the soft crumbly, organic soil, the hoe or cultivator pulls the weeds out more easily. You don't have to wait until after a rain to weed. In a chemicalized soil the weeds in dry weather seem to be be embedded in cement.

25. *Compost is a "safer" material than just ordinary stable manure.* In growing tomatoes, for example, the use of raw manure in many cases results in maximum stalk and plant growth but minimum fruit. Compost is a finished material whereas raw manure has to be worked on by soil organisms to break it down with resultant bad effects on the crop. My impression is that the raw manure as it breaks down supplies an excess of nitrogen in the form of ammonia and the tomato plants respond by excessive vegetative growth and a minimum of reproductive growth, i.e., fruit. Raw manure can also burn plants if applied direct.

26. *Compost kills out weed seeds.* In applying raw manure to the land the farmer and gardener are actually, in many cases, planting a crop of weeds. In the compost heap, the fermenting action kills off most of the weed seeds.

27. *There is less risk of crop failure.* Due to the complications involved in the use of chemical fertilizers, plant diseases, acid conditions

from overuse of chemicals, etc., there is an occasional crop failure. Wind storms blow down stalks of corn and cereal plants such as wheat, oats and barley. This practically never happens where the soil is rich in humus and where plants are accordingly strong and healthy enough to stand on their own legs due to the better root systems they develop.

28. *There is very little plant disease.* In chemical farming and gardening the plant is not nourished properly and thus its resistance is low. Crops raised with plenty of compost seem to be much more immune to disease. Waksman in his book *Humus* says, "Plant deficiency diseases are usually less severe in soils well supplied with organic matter, not only because of the increased vigor of the plants but also because of antagonistic effects of the various soil microorganisms which become more active in the presence of an abundance of organic matter."

29. *The insect menace is reduced to a minimum.* For some reason most insects do not seem to attack healthy plants. Insects are selective in their tastes and those which have not been imported seem to have been trained by Nature to prefer the sickly or imperfectly grown plants. It is Nature's method of doing away with the unfit. Plants grown with chemicals, not being 100% healthy, seem to attract predatory insects.

30. *Few, if any, poison sprays are needed.* On farms and gardens where the organic method is practiced few poison sprays have to be used on vegetables and plants because insect and disease trouble are at a minimum. Large amounts of time and money are thus saved. In heavily infested areas, where compost farming by one or a few farmers only is but an "island" subject to insect invasion on a large scale, sprays may be needed for emergency use, but they can be kept to a minimum.

31. *No chemical treatments are needed for seeds.* Chemicalized farming seems to produce seeds which either harbor or fall easy victims to the organisms of various diseases so that modern agricultural practice demands that all seeds be given a poison bath to kill them off. This practice is not necessary on the organic farm or garden.

32. *It builds health.* In an English school where food was raised by the Howard method there was soon noted a great improvement in the health of the children. In Singapore a similar experiment conducted with a few hundred coolies produced like results. There are many other cases on record which would seem to indicate that the use of humus produces the old-fashioned food rich in vitamins and minerals. There would be a considerable saving in doctors bills if all our food were raised by this method.

33. *Farm animals fed on organically produced feeds are healthier.* Sir Albert Howard fed cattle on feeds raised with humus. He allowed them to rub noses with a neighbor's cattle that had the highly contagious hoof-and-mouth disease. None of his animals caught the disease. There is much proof that the terrible prevalance of diseases of all kinds in farm animals is to a great extent due to their being fed with the devitalized feeds raised with chemical fertilizers. Animal disease is a sure sign of bad farming.

34. *Foods raised organically taste better.* Modern foods grown commercially are losing their old-time taste. Vegetables raised with chemicals are tougher and more fibrous and do not have the full-bodied taste that food has which is raised with compost. Wines in certain sections of France from grapes grown organically are famous for their wonderful taste. Recently in New Zealand, it was discovered that the native cheese tasted better. This was found to be due to the use of less chemical fertilizer under war-time difficulties and restrictions.

35. *The general quality of the crop is much higher.* In England a wealthy farmer who owned many thatched-roof cottages discovered that straw from organically grown crops lasted twice as long as a roof of ordinary straw. In France hay growing in certain sections where practically no chemical fertilizers are used commands a big premium and is shipped long distances for use by race horses. The keeping quality of foods grown organically is far superior. Nowadays when a pumpkin is grown with chemical fertilizers it gets mouldy a few days after it is cut open.

36. *Humus seems to counteract the effects of poisons in the soil.* To quote Waksman's *Humus* again, "The toxicity of plant poisons becomes less severe in a soil high in humus than in a soil deficient in humus; high salt concentrations are less injurious; and aluminum solubility and its specific injurious action are markedly decreased."

There are many more reasons that could be mentioned, some of them quite technical, but those cited show how ridiculous it is for anyone to say that the organic method is impractical because it requires a little extra work to make the compost heaps. This statement is usually made by persons who do not realize all the other advantages and savings in the equation.

While a farmer is applying chemicals to wheat seeds, a thing which would be unnecessary if he were running an organic farm, he could be turning over a compost heap. While he is spraying his fruit orchard he could be making five or ten heaps, depending on the size of the orchard. While he is sitting up with a sick cow he is losing valuable energy which could be used in the compost-making department. While he is fretting in the spring because his land is too wet and cold to start plowing he could be making compost heaps. And so the story goes all the way down the line. Compost farming does not mean more work. In reality it means far less work.

The research department of the School for Living at Suffern, N. Y., under the direction of Ralph Borsodi, made a comparison of costs of buying and using artificial fertilizers as against composts made in conformity with Sir Albert Howard's method. The complete report with elaborate table and annotations was published in the August, 1944, issue of *Organic Gardening* magazine. It is a reliable piece of work which has been thoroughly checked. In it Borsodi says:

"A careful estimate of the cost of fertilizing land for a standard five-year crop rotation of corn, oats, wheat, clover and timothy, indicates that the 'best modern scientific method' costs about $64.39 per acre for the entire five-year period, or on an average about $12.88 per acre per year. An equally detailed study of the cost of using Sir Albert Howard's Indore Process for fertilizing the same land for the same crops, indicates that the total cost can be reduced to $43.92 or to an average of about $8.78 per acre per year. A net saving of fully $4.10 pei year per acre can therefore be effected without taking the superior quality of the Indore Process into account, by abandoning the method of chemical fertilization in general use at present, and adopting this process.'

Bear in mind that the above figures are based on turning the heaps by hand. There will be a marked reduction in costs when machines are used for this work. The mechanization of the compost heap is an accomplished fact, merely awaiting the war's termination for application on a large scale. Mr. Borsodi states further:

"The purchase of chemical fertilizers in large volume began about the year 1880. Yet productivity of farm land has not increased during that time. This fact throws doubt upon the claims for the efficacy of chemical and inorganic fertilizers. The coming of erosion, of dust bowls, or lower water-tables, etc., throws further doubt upon the value of the existing program of fertilization. Finally, the alarming increase in plant and animal diseases and in insect and fungous infestation, raises still more doubts. From the economic standpoint, the cash expenditure—which drives the farmer to raise larger crops in order to secure this cash—is enormously important. Since the farmer's margin of cash returned on each bushel or unit produced is very small, he may have to raise and throw on the market $10 worth of crops in order to get the margin or 'profit' in cash necessary to pay for $1 worth of fertilizer. The farmer, therefore, is doubly handicapped by this outgo—it adds to his expenses, and at the same time it depresses the price received per bushel for what he sells by forcing him to raise and sell many times more in cash crops than he expends for fertilizers. . . . About $200,000,000 has been spent annually in recent years to buy between six and eight million tons of commercial fertilizer materials and mixed fertilizers, exclusive of liming substances and other soil amendments obtained by farmers directly from producers. A general increase in yields per acre, per farm and per man was expected to provide farmers with the cash with which to pay for these chemicals. But no such increases have developed after over a century of mineral fertilization. Farmers as a whole have had, therefore, to increase the acreage cultivated by them in order to pay for this increase in their costs of production.

"If we assume that they have had to raise only $5. worth of cash crops in order to secure a net $1. with which to pay for $1. worth of fertilizers, then they must now be throwing a billion dollars worth of cash crops on the market yearly merely in order to help support the fertilizer industry. It is obvious that the addition of anything like a billion dollars worth of production to the annual surplus of corn, wheat, cotton and other cash crops must play a great part in the constant tendency of agricultural surpluses to depress farm prices. It is also obvious that if there is no real necessity for farmers to buy these chemicals, they would not only save the money and effort put into raising the crops to pay for them, but would also avoid the losses caused by the extra-depressing force exerted on the agricultural prices of all the crops which they raise to sell.

"Experiments and studies made by the School of Living indicate that it is unnecessary for farmers and gardeners to use chemical and commercial fertilizers; that it is possible for them to manufacture all the fertilizers they need on their own farms in order to maintain the full productivity of their soils, and that there are many qualitative benefits which would also accompany the change."

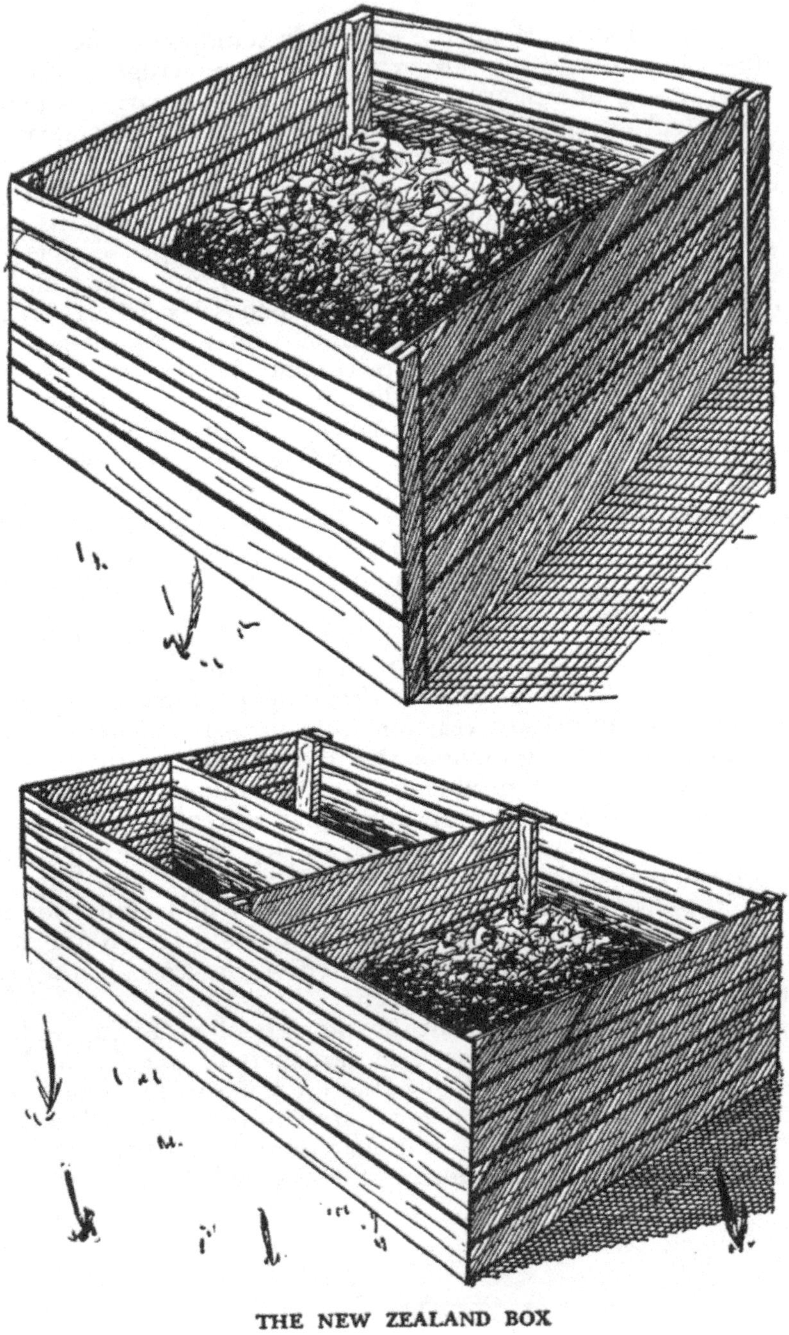

THE NEW ZEALAND BOX

PART THREE

The Dangers of Chemical Fertilizers

I

ARE CHEMICAL FERTILIZERS NECESSARY?

In the 1825 edition of *The Scotsman's Library,* there is an anecdote which is very appropriate to the subject of this chapter:

"A number of years ago, a German quack, who called himself Baron von Haak, vaunted of having discovered a powerful fertilizing manure, which he advertized for sale, pretending that a very small quantity sufficed to fertilize an acre of land, in a very extraordinary manner. Happening to converse with one of his neighbors on this subject, a plain sagacious farmer, the farmer observed to Lord Kames that he had no faith in the Baron's nostrum, as he conceived that the proposed quantity was vastly too small to be of any use.

" 'My good friend,' said Lord Kames, 'such are the wonderful discoveries in science that I should not be surprised if, at some future time, we might be able to carry the manure of an acre of land to the field in our coat pocket.'

" 'Very possibly,' replied the farmer, 'but in that case I suspect you will be able to bring back the crop in your waistcoat pocket.' "

Are the chemical fertilizers necessary in the conduct of the art of husbandry? Prior to the 1850's, chemicals were not used; not ammonium or potassium sulphate or super-phosphate or nitrate of soda or any of the dozens of other strong-acting chemicals. The last figures released by the U. S. Department of Agriculture, showing the use of nitrogen, phosphoric acid and potash by countries, were for the year 1928: Germany used 1,834,000 tons in that year, the U. S. 1,488,000 tons and Russia only 37,000 tons. Egypt uses practically no chemicals and produces the finest crops in the world, thanks to the sediments contained in the yearly overflowing of the Nile.

Since 1928 the use of chemicals on the land in the United States has grown tremendously, and if we may take the figures of W. E. Stokes, writing on the subject of fertilizers in the *Florida Grower* (reprinted in the May 1945 digest magazine, *Your Farm*) close to 10 million tons of fertilizer valued at well over 200 million dollars were produced in 1944 by the more than 800 fertilizer plants of this country. While these figures evidently include *all* fertilizers it is safe to assume that artificials make up the bulk.

Old farmers who remember how their grandfathers grew crops often speak of their method of preserving all organic residues which originated on the farm. They will tell you of the fine crops and of the very little plant and animal disease and insect depredation.

In many sections of Europe agriculture is still carried on with a complete absence of factory-produced fertilizers and those American farmers who have been fortunate enough to witness the work of these peasant farmers never fail to be inspired and influenced by it. Louis Bromfield, the novelist, has farmed in India, France and the United States. On his Malabar Farms at Lucas, Ohio, the use of chemical fertilizers is down to a minimum. Bromfield in the May, 1942, issue of *Country Life,* commented:

> "Let no one ever deceive you: the greatest of soil builders, the most complete of fertilizers is organic animal manure. Fertilizer manufacturers will tell you that it is not a complete fertilizer and other elements not contained in manure or contained in too small quantities must be supplemented, that a corn crop cannot be raised without chemical fertilizers. I am less interested in chemical laboratory analyses than in the evidence of trial by practice; give me all the manure I want, so that I may plough under as much as the plough will cover year after year for three or four years, and I will show you the best crops you have ever seen anywhere."

At our own farm I have taken cheap cattle and fed them on organically produced corn and other foods without purchasing one cent's worth of concentrates and supplements. We have turned out wonderfully finished steers that have brought top prices at the stockyards.

Chemical fertilizers are a form of dope or soil stimulant and can be compared to the taking of medicines. They do not produce basic soil health. Many agricultural authorities more or less admit this and practically every text-book on fertilizers makes apologies in one form or another for their use, although sometimes these must be read between the lines. The soil must have healthy food just like an individual. Many sensible people are now learning to stay away from tonics, vitamin pills and medicines and to depend rather on a varied diet of fresh, natural foods; but too many farmers do not treat their land as a living, vital organism. I should like to quote here from Hermann Rauschning's book, *The Conservative Revolution,* which undertook to analyse the illness of modern European culture:

"Let me say a few words about the ideas of us farmers in this dying age. It has been a period of rationalized farming, a time when men looked on the fields as a retort into which chemical ingredients were put and from which corresponding reactions of amazing harvests were then expected. It was the time of calculation and the treatment of agriculture as a business like any other. We had come under the spell of the rational materialism that placed its stamp on the whole of that superficially successful nineteenth century, with its positivism, its strangely color-blind Darwinism, its complete unconsciousness of the miracle of life with which agriculture had to deal. It was the period of the increase of crops to the utmost limit.

"But we late sons of this rationalist period who have had to learn the final outcome of its teachings have found with horror that this rationalism is a destructive exploitation of living energies. The yields diminished, the soils grew poor. Sulphate of ammonia made the soil sour; the strong dressings of lime ate up the humus; the limit of artificial increase had been reached. On my own farm I had had a field that in the sixties had been assessed as first class for taxation. It was the worst of my fields. My predecessor had allowed it to be entirely denuded of humus by growing successive crops of sugar-beet, dressing with nitrogen, and leaving it without the necessary stable dung. If any dung was given at all, it was poor and lifeless manure, turfed or steeped.

". . . But in the old time only part of the fields was under plow. The rest was sheep run and meadow; the fields that were tilled were given frequent and plentiful manurings with stable dung. An old farm of that type was an organic unity of many individual subdivisions brought into equilibrium through the experience of centuries; each subdivision had its special function in regard to the whole. Each field was assigned only the task to which it was equal, and it was given what it needed for the maintenance of its fertility.

". . . Farming without cattle, without stable manure, this was what the certificated agricultural expert in charge of our experimental farms recommended. And with slate and pencil it could be shown that it did bring in the maximum return. But for how long? And after that, what then? But who cared for that aftertime? Had not everyone to take thought for the moment, for himself only? If the yield fell, could not the farm be sold? Could not the same process be successfully continued elsewhere, by 'organized exploitation'?

"I will not weary you further with this subject, but it was just the thinking men among us farmers who began to be uneasy about this humbug of progress and rationalization and to think once more of the farming methods of their forefathers. . . .

"But I should like to say this: everything that is true of farming may be equally true of the life of a people. 'Organized Exploitation'—that is the phrase with which a later time will sum up that superficially grandiose and successful nineteenth century. Organized exploitation is the hidden fault in all the efforts of the doctrinaires and utopians who want to wrest man by force out of his own nature in order to deliver him over to the tyranny of absolute reason. That party of rational progress, of rational planning, of a new order of society trained in accordance with the dicta of doctrinaires: all that is of yesterday, and its sterility has been proved by these crises. It is all nineteenth century, all of the past, not the future. They are all out of date, all these men who are up in arms against the nature of things and against an order which is ordained of God. There is no pursuing this path without final destruction. Ahead of us on this path lies, both literally and figuratively, the destiny which an agricultural expert forecast in an inquiry some years ago—the conversion of Europe into a steppe."

Plants growing in highly chemicalized soil do not have natural resistance to disease, just as people who do not have bodily resistance fall prey to disease more quickly than the average person. Then it becomes necessary to spray poisons on these "sick" plants to protect them against all manner of disease and insects. The use of these sprays, which are absorbed into the ground, further pollutes an already contaminated soil. Between the chemicals and the spray poisons, the biologic life of the soil is reduced almost to the vanishing point. Gone are the earthworms, and the bacteria and the fungi almost disappear; the soil is practically dead. The farmer might just as well farm in sand loaded down with chemicals. Dr. Ehrenfreid Pfeiffer mentions vineyards he has seen which, because of frequent applications of copper and lime sprays, do not show a single earthworm, and have thereby lost a valuable ally.

The average farmer who uses chemical fertilizers is actuated by several motives. He is after larger yields, wants his crops to mature earlier and is trying to save labor. He has been led to believe that this is the height of scientific farming. The fact that just as large, and larger, yields may be obtained with compost, and were, in his grandfather's time, is overlooked. An important point frequently ignored is that the *quality* of the crop is far superior where organic fertilizers are used exclusively.

In *An Agricultural Testament,* Howard describes a project in Indore, India, where humus was used exclusively on a cotton crop. The average cotton yield in surrounding country was 200 lbs. per acre. In the first year of the experiment (1927) the average yield on the experimental farm was 340 lbs. per acre. In 1928 it increased to 510. In 1929 it reached 578. A significant fact revealed by this experiment was that yields were maintained despite reductions in rainfalls which seriously affected the surrounding farmers' crops.

In the autumn of 1943 the Horticultural School of Ambler, Pa., became interested in experimenting with the Howard method on their 100-acre farm. By the spring of 1944 they had made 100 tons of compost which they used in growing oats and corn. In each case, compost was applied on one-third of the field, raw manure on one-third and raw manure plus chemical fertilizers on the balance. The results in favor of the compost were so noticeable that they could easily be seen in the field with the naked eye. They now sell compost made by the Howard process to many small gardeners.

In a patch of our own, we grew about fifty sunflower plants with compost, using a larger amount of humus than usual because we want the seeds as delectable tid-bits during the cold winter when we don't have either sun or fresh-grown vegetables. The plants are gigantic, fifteen feet tall—a nine-day's wonder in Pennsylvania. The stalks are about as thick as a man's fist. Passers-by stop and gape. The yield is not only more than double the average, but each seed is large, making it worthwhile going through the action of cracking it open. But most important of all, their taste as compared with store-bought seeds is as champagne to ginger-ale.

The Chinese get enormous yields by using composts and other organic materials. In *Farmers of Forty Centuries,* F. H. King describes yields of wheat in China of over 100 bushels an acre as not uncommon. The average in Szechwan province was about 25 bushels per acre. In this country our average is now about 18 bushels. King's book (unfortunately out of print) contains hundreds of photographs, many of which show the multiplicity of methods used by the Chinese in making composts and in using other organic materials. Of course the Chinese farm small acreage but there is no reason why we cannot apply some of their methods on large acreage by developing the proper machinery.

When thinking about China there always comes to mind the joke about the fertilizer salesman who was sent to the Celestial Empire but who apparently didn't stir up a single order in months. When he received a cable demanding when they could expect some business he replied, "Trade impossible, with 400 million competitors."

Some farmers employ strong chemical fertilizers to speed up the maturity of the crop so as to get to market early and enjoy the premium paid before the market becomes glutted and also to clear the land for a second crop. Nitrate of soda is often used by truck farmers for this purpose. There are reasons—amplified later—why many of us consider that food produced in this manner lacks essential nourishment.

Do these farmers actually save time by using such artificial methods? By going against nature they suffer later in the form of plant diseases and insect depredations. They have to spray poisons and ruin their soil so that it is no longer a healthy place, from a biological standpoint, to raise crops. They have to spend money for such poisons, must invest in expensive spraying equipment; must spend much time in spraying, and as the situation gets worse and worse each year, they worry themselves sick about it, which shows that the Bible's "as ye sow, so shall ye reap" smites them in an unexpected way.

A very important reason for not using chemical fertilizers is the fact that their use makes the plant gradually lose its reproductive power. The seeds lose their potency, and the variety runs down; the species dies out. Howard has stated in *Organic Gardening:*

> "Practical experience has already shown that the proposed combination of humus and artificials is inadmissible. It leads to disaster in two directions; (1) the variety soon loses the power to reproduce itself—in other words, it runs out, and (2) the crop after a few generations is unable to resist disease. Two crops—the vine and sugar-cane—which have been cultivated for hundreds of years—supply the evidence. In the Orient these two crops, as grown by the cultivators, are always manured with farmyard manure; artificials are unknown. There has been no change in the varieties grown for many centuries; they are to all intents and purposes eternal. But in the West, where the manuring is for the most part by means of artificials, the story is a very different one. In both crops the varieties are constantly running out and plant breeders are called upon to produce a never-ending stream of new kinds. The use of artificials therefore destroys the power of both these crops to reproduce themselves.
>
> "How can artificials alter the constitution of plants and initiate this running out process? I suggest the answer will be found in the gradual reduction of the fertility principle carried into the plant from a soil in good heart by means of mycorrhizal association. This association is the means by which fungus threads, rich in protein, are digested by the roots and carried in the sap to the green leaves. As humus is replaced by artificials, this agency in the feeding of plants gradually disappears and as it does so the capacity of the plant to reproduce itself gradually fails. In other words, the running-out process is a natural consequence of methods of manuring which have failed to pass the test of time. What seems to matter in the stability of the variety is the circulation of protein from soil to plant and animal, and back again to the soil by way of properly constituted humus. If we interpose a substitution phase in Nature's round, we weaken one of the links in the wheel of life and a breakdown soon follows."

Certain varieties of grapes that have been growing for centuries in India and Persia still retain their reproductive potency whereas in southern France new varieties have to be introduced constantly to replace the old. At the Rothamsted Agricultural Experiment Station in England experiments were begun in about 1843, which lasted for more than 50 years, to determine whether wheat could be grown continually with the use of chemical fertilizers only, that is, without rotating crops. While the yields showed up favorably for the entire period, it was not generally known that new seeds were brought in each year from the outside. If the seeds from these continuous crops had been used, *that* would have been a test and it would undoubtedly have drastically affected the yields in later years.

Where organic fertilizers are used in the form of green matter only, without the aid of animal matter, a similar problem of the running out of the variety seems to occur. Howard has found that animal matter in the form of stable manure or in other forms such as fish and chicken cuttings, must be used.

F. C. King, in his *The Compost Gardener,* gives his experience as a practical gardener:

"Let us examine the facts relating to the use of sulphate of potash on tomatoes. I unhesitatingly declare that no commercial grower of tomatoes would dream of using sulphate of potash on tomato plants if he intended to save seed from those plants for his own use. He knows full well that in a few years' time, if artificials of any kind are used, his strain of plants would be worthless.

"A simple test will soon prove my assertions. Grow half a dozen tomato plants either in pots or boxes filled with compost, particularly using old tomato stalks and leaves in the heap, and compare them for flavor with any quantity grown with artificials. You will not be disappointed either in weight of crop or flavor. Save your own seed from the half dozen plants and get even better results the following year. Then try saving seed from plants fed on artificials. There will be a marked difference in the third season; in fact, with the seed saved from these plants you will have difficulty in raising a stock after three years. If we compare the time spent in preparing compost for tomatoes with the time saved by using chemical manures and judge the results fairly, we are better able to assess the merits of the two systems. I have tried both and if, by the labour-saving use of artificials, I could produce crops equal to my compost-grown vegetables I would have continued the use of artificials."

In many cases when a variety is beginning to give trouble and in its death throes starts to throw fits and begins to act temperamental, the experts look around and by sleight-of-hand produce a new disease-resistant variety which is announced with a blare of drums as the sensation of modern scientific agriculture. But if you will look behind the curtains you will in many cases see that the process is not as involved or as scientific as one might imagine. The scientist starts with a clean slate. He will take a variety to start with that comes from another country where more wholesome methods of fertilization prevail and he will cross it with some other varieties. He knows that what he calls "primitive" plants are stronger than our run-down varieties.

Let us take, for example, the case of Texas cantaloupes which were suffering badly from downy mildew and from aphids. Dr. S. S. Ivanoff of the Texas Agricultural Experiment Station developed a new variety resistant to both mildew and aphids. According to the May 13, 1944, issue of *Science News Letter:*

"Breeding stocks were selected from four varieties of West Indian origin, all of which had shown good resistance to aphids and mildew under South Texas conditions. Good size and shape for market requirements, ability to stand up under shipping conditions, and desirable qualities of sweetness and flavor were developed during the breeding program."

Please note that Dr. Ivanoff started with a West Indian product and that all the good characteristics of these West Indian cantaloupes are known to be present in crops grown with compost.

How many times do we read that trouble is being encountered with local varieties of wheat and that our scientists are sent to Russia and Central Asia to introduce healthier varieties from there, ones that do not wilt as soon as the wind blows from the wrong direction. This is interestingly shown in a letter that Dr. Miron Taborowsky wrote to the *New York Times* in February, 1944, in which he says in part:

"Carleton, a leading agriculturist, had an opportunity to observe the bad effects of the drought and black rust, which were killing the wheat in Kansas for many years, causing disaster to the farmers. Neither affected the wheat which the Russians grew from the seed imported from their native country. Inspired by this idea, Carleton went twice to Russia to import seeds of the Kubanka wheat and later of the Kharkov wheat. The wheat took successfully in the soil of the

United States and by 1914 half the winter wheat of the country was of the Kubanka and Kharkov variety."

As we have seen in the first part of this chapter Russia prior to 1928 used practically no chemical fertilizers.

2
CHEMICAL FERTILIZERS ARE TOO STRONG

That chemical fertilizers are poisonous is a well-known fact. One of the most frequent offenders is sulphur in its many forms. Sulphuric acid has been used often as a suicidal agent, where difficulty was experienced in obtaining other poisons. It causes disfiguring scars if allowed to come in contact with the skin. Yet it is a very important element in the manufacture of one of our most popular chemical fertilizers—superphosphate. According to Professor G. H. Collings in *Commercial Fertilizers,* in 1939 out of a total of 1,994,023 tons of fertilizers used in the United States, 702,135 tons were superphosphate. It is made by taking raw, ground phosphate rock and putting it into a chemical action with sulphuric acid.

Most agricultural colleges recommend the daily spreading of superphosphate in the manure of cattle barns to act as a deodorant and to "lock in" the nitrogen of the manure. Yet this practice sometimes causes a disease in cattle which actually eats away part of their hoofs, (not to be confused with hoof-and-mouth disease).

Dr. C. E. Greene in his challenging book, *The Cancer Problem,* goes into detail as to the effects of sulphur in various industries producing cancer. For example the chimney-sweep, who is constantly in contact with sulphur in coal soot, headed the list of cancer sufferers, dying from this disease (1900-1902 figures) at the rate of 136 per 100,000 population. Tanners were at the bottom of the list with 33 deaths from cancer per 100,000. On the other hand furriers rated 111 per 100,000. Dr. Greene states:

> "The tanner steeps his hides in lime, which is apparently a prophylactic, while the furrier adopts a different process. The skins of mammals, we are told, are best preserved by a mixture of four parts burnt alum to one part of saltpetre. Alum, it must be remembered, is composed of sulphuric acid, alumina, an alkali, and water. Nearly all furs have also to be dyed, and the mordant used is chiefly sulphuric acid. Most of the cheaper furs are cleaned with benzene, which is a coal-tar product agitated with strong sulphuric acid. They are dyed with aniline dyes of the coal series, which are constantly produced by heating with sulphur."

Dr. Greene shows a list of about 58 occupations and it is most uncanny how the incidence of cancer deaths parallels the use or contact with sulphur in the various industries. I have tried to secure for comparative purposes modern American cancer statistics by industries, but they do not seem to be available.

Cancer seems to be most prevalent in cold countries. "The disease increases in frequency with remoteness from the equator to such an extent that in latitude 60° N the rate per 100,000 of the population is 105.7 in cities, whereas in latitude 10° N to 10° S the rate in representative cities is only 40.9." Dr. Greene explains this to his satisfaction by showing that in the north coal (which naturally contains sulphur) is used for heating, and in the south very little is used. Other experiments have shown that coal-tar compounds can produce cancer in mice.

He made a detailed study of the cancer statistics for every city, village and hamlet in France. It is remarkable how closely it follows the use of coal as a fuel. In a section where coal is the predominant fuel, cancer is higher than where wood is used.. Cancer in the northern section of France is much more frequent than in the south. Further north more fuel has to be burned. "In Nairnshire, Scotland," says Dr. Greene, "cancer seems to be actually non-existent in the districts where nothing but a light non-sulphurous peat is burned, while it is very common in that part of the country where coal is used. In Orkney, again, cancer is common in a few districts where nothing but peat is burned when that peat is coal-like in character and has a very high sulphur percentage."

Dr. Greene noticed that some parts of London have a death-date from cancer consistently ten times greater than that of others. When he checked it, he attributed it to the unevenness of the roofline of the houses. In other words, where all the houses are of about the same height the smoke can get away easily, but where there is a great variation of the roofline, whirlpools and eddies of smoke occur which send it into certain homes and the cancer rate goes up.

Dr. Greene mentions the city of Erfurt, Germany, where the statistics show that professional gardeners suffer much more from skin cancers than the rest of the population. He attributes it to handling infected earth. Germany was the originator of the whole chemical fertilizer idea, and much more artificial fertilizer has been used there in the past than in the United States.

This reminds me of an unusual item in *The Land,*[1] "The Dirt Beneath Our Feet," by Dr. Karl Menninger, who writes, "Belatedly we have discovered, of course, that a tremendous anti-bactericidal element is contained in the soil. When I was a child and worked on the farm and my uncle would cut his fingers and rub them in the dirt, I would tell my father. He was a physician, and he would shake his head and look half disgustedly and think, 'It is only a question of time until I have to amputate my brother-in-law's arm.' He never did amputate that arm and his brother-in-law got along all right with his wound sterilized by dirt, which I was religiously taught in medical school to take away from the patient immediately, regardless of how much pain it caused the individual to be scrubbed."

A discussion of sulphur compounds should include coal ashes. Many old-time gardeners swear by their use as a soil amendment, not as a fertilizer but to aerate heavy soils and improve their physical condition. Most authorities state that soft-coal ashes should never be used as they are toxic to plants, but many do sing the praises of the ashes

[1] Vol. 3, no. 2, 1944.

of hard-coal. Actually it is merely a matter of degree. Soft-coal ashes contain more sulphur. We do not recommend the use of either. If there is poor combustion in the furnace the ashes may then contain tar oils, sulphites and metal oxides which are detrimental to soil life: The railroads strew soft-coal ashes along their rights-of-way to destroy vegetation. Notice how effective they are.

In May 17, 1944, a bill was introduced into Congress to put the regulation of the manufacture of fertilizers under the jurisdiction of the U. S. Department of Agriculture. Under this bill it would not be permitted to ship fertilizers that contained coal cinders in inter-state commerce.

The organic gardener has no need for coal ashes to aerate his soil. By means of compost he will soon build a spongy soil structure which will provide fine aeration. It might take several seasons but it will pay to wait. I cannot conceive of anyone who loves the soil polluting his garden with coarse ashes. Aeration can be achieved with peat moss or even sand. It might cost a little money, but the garden will benefit by having a rich, dark, soft, smooth soil which you will love to run your hands through.

Not only sulphur, but potassium too may be harmful in fertilizers. F. C. King, in a recent article in *Organic Gardening,* draws attention to the poisonous nature of some potassic fertilizers:

> "I should like to say a little about potassic manures. The chief source of supply of potash salts is Germany. These are marketed as kainit, muriate of potash, sulphate of potash, etc., but as they all come from the same mine, though they vary considerably in price, there is little to choose between them. The difference in price is merely a question of the quantity of impurities they contain. The higher the price per hundredweight, the less the impurities, but at best, as in the case of muriate and sulphate, they contain more than 50 per cent impurities, while in the case of kainit the percentage of waste is 88 per cent. It should be clearly understood that these impurities are very detrimental to plant life and are, in fact, definitely poisonous in most cases. For the most part they are too dangerous for amateurs to use, although forms of potash are constantly being recommended for such crops as onions and tomatoes."

In a recent issue of the *Eastern States Cooperator* appears the following:

"Almost every nitrogen carrier has some disadvantages. The more soluble carriers all have some tendency to pick up moisture and become difficult to distribute in rainy or humid weather. Some of them are considered a fire hazard. No mineral nitrogen carrier should be left around carelessly where livestock might nose into it because it is toxic to animals."

Ammonium nitrate is a chemical fertilizer which will explode if it receives a strong detonation. While the chances of an explosion are one in a million, I wouldn't want to have stuff like that around my barn. Bags in which this material has been stored are inflammable and the manufacturers usually advise that they should be destroyed as soon as emptied.

There are rare trace elements in the soil which Nature, the perfect laboratory technician, knows how to handle. They include boron, fluorine, silicon, iodine, cobalt, chlorine and others. These elements seem to be needed in extremely small quantities, an oversupply being toxic to plants and animals. Where organic fertilizers are used there never is a question of trace elements lacking and crops rarely show any of the deficiencies which the practiced eye can note where boron, for example, is absent. Of late, chemical fertilizer manufacturers have been experimenting with these trace elements, adding them here and there to their products, but in view of the danger of an over supply causing toxicity, a farmer should think twice before using them in that form.

A simple example of how a chemical will inhibit plant growth is the case of sea-spray, milder than many of our chemical fertilizers. I quote from the *Bulletin* of the Torrey Botany Club:

"The tendency of many woody plants on seacoasts to show a sloping form of growth with the main stem pointing at an angle away from the sea has previously been thought to be due to the force of the prevailing winds—hence the term generally applied to the form of such trees, 'windform.' Wells and Shunk of the North Carolina State College of Agriculture and University of North Carolina, show that, at least on seacoast areas, this is not due to the force of the wind but of sea-spray deposited upon the young shoots on the seaward side of the plant; this spray prevents growth on that side of the tree. The effect is therefore not physical in nature but chemical, and the term 'wind-form' should be replaced by 'spray-form.' "

Sir Albert Howard comments on this:

"In the South of England, and also in the Northwest (in Westmoreland) we sometimes experience heavy gales after the trees come into leaf in the late spring. These exceptional gales are loaded with salt spray. The damage to the young leaves must be seen to be believed. It just burns them up."

3

CONDEMNED FARM LANDS—CHEMURGY

Up to now we have dwelt upon the effects caused by man, who, by his own hands, pollutes the soil. There are other cases where the soil is fouled or poisoned indirectly by man, or by certain natural geological occurrences in the soil formation. The adverse effects of sulphur, which have been discussed, occur chiefly in industrial areas where the air is especially corrosive due to the large amount of sulphur in factory smoke. The following item appeared recently in *Horticulture Magazine:*

"The Victory Garden Committee for the Chicago Metropolitan area recommends the following vegetables to be grown in smoky and congested districts: Carrots, kale, turnips, beets, collards, Swiss Chard, mustard greens, leaf lettuce or Cos lettuce, radishes, Spring onions (from sets), endive and parsley.

"Metropolitan areas are sometimes affected by smoke from nearby industrial districts. This smoke contains sulphur dioxide gas which acts as a poison to some plants and causes their leaves to wilt and their blossoms to shrivel and fall off. This happens most noticeably on quiet damp days when the smoke and gases settle near the ground like a pall. Many other cities are, of course, in the same position as Chicago."

In purchasing a farm one should be careful to avoid places that are polluted by the smoke of railroad trains which burn soft coal. Land near certain factories should also be by-passed. Bear, in *Soils and Fertilizers,* says, "marked injury to vegetation has been noted in the vicinity of cement plants as a result of the large amounts of soluble potash in the dust that is deposited on nearby vegetation. Similarly, a large part of the plant life has been destroyed over large areas of land surrounding smelters in which sulfide ores were being roasted." Here again we note the caustic effects of sulphur. In the *Veterinary Journal,*[1] Norman Bisset describes a chronic copper poisoning in sheep grazing near a nickel-smelting factory.

In U. S. Department of Agriculture *Miscellaneous Publication No. 369,* "The Mineral Composition of Crops," a rather astounding statement appears:

"The problem is intensified because of the location of our large cities and of our truck farms that supply these cities with food. Most of our large cities are located on the well-leached podzols and podzolic soils that are acid in reaction and low in the bases. The truck farms are of necessity located near the cities on the same soils, and large quantities of truck crops and fruits are shipped to these cities from the localities where nutritional disorders due to mineral deficiencies in the soils have been noted in both humans and animals.

"Although many factors other than the poor quality of food probably are responsible for dietary difficulties, this factor is believed to be an important one; and if it is important, then it certainly remains the duty of the agriculturalist to supply, for at least the protective portion of the diet, a food of the highest nutritional quality."

But almost no attention is being paid by farmers, agronomists, physicians or the general public to such plain-spoken admonitions as this. Our food crops are certainly being raised in a most slipshod manner.

[1] Vol. 90, pp. 405-407, 1934.

In South Dakota, Wyoming and Nebraska and many more states in the far west, selenium, an element similar to sulphur in its characteristics, is found in the soil. It is poisonous to livestock. This element seems to be found in geological formations associated with prehistoric volcanic eruptions. In the January, 1943, issue of the *Scientific American,* J. V. Shepard recommends, with respect to this problem, "Highly toxic areas may either be fenced off or planted with non-food crops which are to be used in industrial products." In one section of South Dakota the government has already withdrawn 100,000 acres from cultivation. For a long time the Government as well as economists and sociologists have been making studies of sub-normal acreage which has been designated as marginal land, but no basic principle has as yet been arrived at as a guide in the program.

In many states the soils in fruit orchards, due to heavy spraying with lead arsenate, have become so toxic that nothing will grow there any more, not even weeds. In the state of Washington alone there are about 15,000 such acres lying idle. The Agricultural Experiment Station at Pullman, Washington recently issued Bulletin No. 437 (March, 1944), *Vegetable and Small Fruit Growing in Toxic Ex-Orchard Soils of Central Washington.* In it there is outlined a plan for restoring such land to be used for growing vegetables and small fruit, by a process of using chemical fertilizers, cow manure and plowed under cover crops. This booklet and its ill-contrived plan shows a certain disregard for the safety of public health. That side of the problem is not discussed. There is plenty of evidence, according to Professor Firman E. Bear of Rutgers University, "that plants growing on soils that have been treated with arsenates contain more arsenic than do the same species when grown on untreated soils." Lands such as this might better be condemned as a food-raising medium, designated as sub-marginal and used for producing industrial crops only.

Chemurgy

A new science has recently grown into being concerned with seeking uses for crops in industry. It is called *chemurgy.* Here is a typical "chemurgic" advertisement—"Tomorrow a Motor Car Body, a Square Meal and a Suit of Clothes from the same crop!"

Much has been written about Chemurgy in its search for new uses for farm or forestry crops in industry. There has been formed a "National Farm Chemurgic Council, Inc.," with offices at Columbus, Ohio. A booklet issued by them states that their mission is "to advance the industrial use of American farm products through applied science." Their booklet further states that "Farm Chemurgy," liberally translated, means putting chemistry and related sciences to work in industry for the farmer, and indirectly, for society in general. There is even a magazine, *Chemurgic Digest,* devoted to these industrial crops.

In the long run this new field of endeavor, if not curbed or controlled may prove harmful to society, for it may extend the single-crop technique of land-mining with all its attendant evils of soil exhaustion and erosion.

The Chrysler Corporation alone is using farm products in making automobiles to an extent which requires 13,350,000 acres a year. These products include wax, fats, animal hair, wool, leather, lumber, cellulose, vegetable oils, glycerine, gums, starches, fibers and others. Even the Government, in its various regional experimental laboratories, seems to be going all-out to find new industrial uses for farm products. One set of Government scientists is at work turning corn stalks into paper. Another is making glue out of tapioca, or combustible gases from corn cobs. Four Government laboratories, employing many hundreds of researchers, are exploring the field.

C. C. and S. M. Furnas in *The Story of Man and his Food* point out, "We do not only deplete our soil for food removal but for textiles and raw materials for chemical industry. These are worthy advances of civilization, but they are dangerous practices for the nutrition of the future *unless the agrobiologists learn the trick of completely re-fertilizing the soil.*" (Italics mine.) This is a vital point, so vital that, considering the erosion and exhaustion problems we already face as a nation, we cannot possibly overemphasize it. In the interests of the food-consuming public a survey should be made right after the war. Those lands which for one reason or another are not fit for growing food crops should be designated as sub-marginal, and for use in chemurgy. In some cases these industrial crops grow naturally on lands which cannot grow food crops on a practical basis. For example, in Michigan, milkweed is being grown on a large tract of cut-over timber land. The soil here is too sandy for food-crop cultivation. Cattails are being grown in swamps and at the sides of ponds. Reforestation, on a vast scale and carefully regulated, can supply a large chemurgic need for wood-cellulose for all the varied uses.

Some of the new products that stem from the soil are highly ingenious. There is a non-metal can made from a plastic material, the ingredients of which include cereals, linseed, animal tissue and bones. One company has a factory for converting corn into chemicals used in photographic films, motor fuels, liquid soap, embalming fluids, and many other products. This corporation converts 11,000,000 bushels of corn annually. There is an automobile lacquer made from potatoes, gasoline from pumpkin vines, and so on to the tune of many millions of acres devoted to non-food crops.

Does the average farmer have the training that enables him to act as a true custodian of the nation's land? Our land, actually, is the basic capital of the nation and should be used in that spirit. This is a free country, we all admit, but does freedom mean the wanton dissipation by anyone, who so chooses, of our limited store of soil fertility?

We are beginning to realize how much of our capital we have lost in the uncontrolled practice of denuding our forest acres, leaving nothing behind but brush and slash, for forest-fire and erosion to work upon and create deserts. We are beginning to realize the disastrous nature of farming malpractice that creates erosion and dust-bowls. From Washington's time on it has been clear that single-cropping destroys the fertility of the soil and eventually makes it "not worth plowing"—but even to date in many areas of the Nation little has been done to change the practice: we are still destroying the land and moving on, leaving the barren ground behind.

If Chemurgy is to accelerate this process—make single-cropping more profitable for a time, as a handmaid to industry, only to result in vaster dustbowls and more extensive marginal lands—Tobacco Roads—then we shall have taken a long step towards making our good earth another Sahara Desert. That, too, was a fruitful country once, when it was wooded and cultivated property. So was the Gobi Desert, and the Valley of the Yangtze.

4

WEED KILLERS AND DISINFECTANTS

Besides chemical fertilizers, there are other strong poisons that come into play on the modern farm. For example, weed-killers. Poisons such as arsenate of lead, sodium arsenite, arsenic acid, ammonium thiocyanate, acidified copper sulphate and many others are used. A few years ago the San Bernardino Humane Society of California got up in arms because chemicals, which were applied around electric power-line towers to destroy vegetation, were killing off deer and domestic livestock.

Farmers often use these strong chemical weed-killers, without knowing that they kill his best friends, the soil bacteria, fungi and earthworms. And when professors of agriculture recommend their use on a large scale it is sad indeed. A standard reference work, *Weed Control* by Robbins, Crafts and Raynor shows pictures and recommends converted orchard spray rigs for this purpose. These have as many as thirteen nozzles that can discharge 10 gallons a minute and which have sprayed as high as a thousand acres in one season.

Here is an item from a publicity release of the University of California, dated June 30, 1943: "Celery is the latest crop to be weeded by selective sprays, it is reported by R. N. Raynor, associate in botany on the Davis campus of the University of California. Celery beds, from which plants are transplanted to the fields, have been weeded by hand, formerly, at a cost of up to $300 an acre. Stove-oil spray, at a cost of $10 an acre, does the job satisfactorily, it is said." What stove oil will do to the soil, its earthworm population, and indirectly to the consumer's stomach is as nothing compared to the dollars saved by the truck-farmer, although in the end he will actually pay over and over again in dollars for ruining his soil.

Only a few months ago a use of sodium chlorate to control bindweed in California boomeranged and caused to appear a mosaic disease in tomato fields. The yields were drastically reduced. This was discussed by Dr. John T. Middleton of the California Experiment Station at Riverside California.

In a great many cases "weed-infested" lands come about because of artificial "fertility" due to the use of chemical fertilizers. In addition to making the land acid, and creating hard-pans below the surface, as we have already noted, these fertilizers create changes in the soil structure which encourage certain types of weeds to flourish. To give you an idea of this principle, in bombed and burned areas of London, over 100 species of strange plants never before seen by present-day Londoners, have recently been discovered. In the ruins of the Great London fire of 1666 there appeared for the first time in 270 years a plant called the London Rocket. Botanists are watching for its reappearance today.

A worn-out soil encourages certain weeds, including sorrel, red top, daisies, purslane, hawkweed, dewberries, and broom sedge. In fact broom sedge flourishes quite profusely when a farm is almost ready to be abandoned. By studying the habits and habitats of weeds you can become expert in judging the condition and fertility of many soils.

Sir Albert Howard in his *Testament* describes a remarkable case where a growth of mares' tail indicated a badly aerated subsoil:

"One of the most interesting cases of pan formation that I have observed in Great Britain was on the permanent manurial plots of the Woburn Experiment Station, where an attempt to grow cereals year after year on the greens by means of an artificial manure has been followed by complete failure of the crop. The soil has gone on permanent strike. The destruction of the earthworm population by the regular application of chemicals had deprived the land of its natural aerating agencies. Failure to renew the organic matter by a suitable rotation had resulted in a soil devoid of even a trace of tilth. About 9 inches below the surface, a definite pan (made up of sand particles loosely cemented together) occurred, which had so altered the aeration of the sub-soil that the whole of these experimental plots were covered with a dense growth of mares' tail (*Equisetum arvense L.*), a perennial weed which always indicates a badly aerated sub-soil.

Nature as usual had summed up the position in her own inimitable fashion."

Another case noticed by Howard, showing that certain semi-parasitic weeds can be eliminated by enriching the soil with humus, is mentioned in his book:

"One of the pests of maize in Rhodesia—the flowering parasite known locally as the witch-weed (*Striga lutea*)—can be controlled by humus. This interesting discovery was made by Timson, whose results were published in the *Rhodesia Agricultural Journal* of October, 1939. Humus made from the soiled bedding in a cattle kraal, applied at the rate of 10 tons to the acre to land severely infested with witch-weed, was followed by an excellent crop of maize practically free from this parasite. The control plot alongside was a red carpet of this pest. A second crop of maize was then grown on the same land. Again it was free from witch-weed. This parasite promises to prove a valuable censor for indicating whether or not the maize soils of Rhodesia are fertile. If witch-weed appears, the land needs humus; if it is absent, the soil contains sufficient organic matter. Good farming will therefore provide an automatic method of control."

An interesting illustration of the same point is shown in a communication received from Bernice Warner of the Ohio Earthworm Farms:

"Last fall we failed to notice any crab grass. There wasn't any, where before it had been terrible; but it took us about a three-year fight. We put earthworm castings on the spots that had it (therefore feeding it well) and cut all the grass 'high' to give the good grass a chance to get better root construction."

Besides building an organically fertile soil to keep down undesirable weeds, there are other safe methods, such as the proper rotation of crops. A particular crop may always be associated with a certain weed that has an affinity for it. Oats, for example, seem to encourage the growth of wild mustard. If the same crop is grown year after year the weed tends to increase. Another means is the growing of "smother" crops such as buckwheat, alfalfa, rape, sorghum, millet, barley, sunflower, the clovers, ensilage corn, etc. For various reasons these crops crowd out, or smother weeds, and are therefore important in weed eradication.

Weeds are best defined as plants for which man has not yet found a use. Our lands will never be entirely free of them; you wouldn't want them to be. Weeds in moderation serve a valuable purpose. They capture food elements from the soil that might leach away were the ground left bare. When they decay these are returned to the soil. Their roots furnish valuable organic matter to the soil when they die. Living, healthy soil may be compared to a factory; the workers (bacteria, fungi and earthworms) are constantly converting raw materials like humus from insoluble to soluble plant foods. These can best be stored up in weeds when not being utilized by crops.

Weeds too can act as sub-soilers aerating hard pans through their root activities and keeping idle crop land porous and workable.

Chemical Treatment of Seeds

Another practice on the farm, which has become very popular in recent years, is the chemical treatment of seed to prevent seed-borne diseases such as smut. Naturally, poisonous material has to be used such as formaldehyde, copper and mercury dust, strong enough to kill off organisms that cause disease. Dr. Ehrenfreid Pfeiffer has told me that these poisons are absorbed into the seed and may be found later in the plant. Other authorities do not consider this a dangerous practice, since the amount of chemical absorbed would be slight. Now that it is possible to detect the presence of things like copper by means of the spectroscope, work should be started to see definitely what happens to these poisons, and if they occur in the crop.

Professor Lutman of the University of Vermont discussing the problem of smut, writes:

"It could be prevented by the use of seed from clean fields only, or by carefully eliminating all smutted heads in a seed plot, but the average farmer would not do it. So he falls back on chemical treatments. I wonder if a thorough *washing* of the seed grain would not reduce the smut?—I can't recall any experiments along that line—but the smut spores are only external and a thorough scrubbing ought to reduce the numbers, certainly."

To show how dangerous to handle these seed-treating chemicals are, witness an admonition given as to their use in the May, 1942, issue of the *Macdonald College Journal*, "Seed Treatment," by R. A. Ludwig:

> "The chief disadvantage of the copper and mercury compounds is their poisonous nature both to humans and to livestock. In handling these dusts the precautions recommended by the advertisers should be strictly observed. While making the treatments the mask, obtainable with the mercury dust, should be worn. Left-over seed grain should not be fed to livestock and containers holding the mercury should be kept out of reach of children."

In applying these treatments the seed may be injured and the result may be feeble plants—which easily fall prey to disease. These poisonous compounds are complicated to handle and may not be used promiscuously for every kind of seed. Inexperienced farmers are apt to damage costly seed through bungling the instructions.

Chemical seed treatments are needed only by those who engage in an extensive chemicalized form of farming. When chemical fertilizers are used as well as poison sprays, a vicious cycle is created. The seed must be chemicalized, the land chemicalized and the crops must be sprayed from start to finish—even with poisonous gases like cyanogen in some cases. It is a completely chemicalized process from beginning to end. If seeds are grown in a healthy soil full of humus such chemical measures are unnecessary. We do not employ any on our farm, and have no need of them.

The unfortunate part of the whole business is that growers of seeds are now taking matters into their own hands. Because of the average farmer's and gardener's ineptitude in handling these chemical seed treatments, some seed houses furnish seed already treated. When you buy seed corn it is already blackened with a chemical. When you get a sack of wheat seed it carries the admonition that it must not be fed to cattle. Many ordinary vegetable seeds, in their innocent and colorful looking packets, are also being dosed with these chemical poisons. In Great Britain, Howard informs me, some seed growers insist on their seed being grown solely with humus, but so far as I know there are no such seed growers in the United States.

There is another practice that should be mentioned here. Wheat is usually stored in big piles in the barn and in order to prevent its infestation with insects, the piles are sprayed with cyanogen gas. This is a very strong poison and the farmer has to be protected with a mask while working. The trouble is that the farmer is not aware of the consequences of all this chemicalization. To him it is merely one of the routine, incidental procedures of scientific farming. If he becomes conscious of his obligation to himself and the public he would work out ingenious methods to store all his crops in the safest manner possible. Sun-drying, for example, is an excellent way of preservation since pests prefer to attack damp seed. The first season on the farm we sprayed the wheat pile with cyanogen, but a few years ago I adopted a method of storing the grain in shallow boxes with square holes on the bottom covered by a wire mesh to permit air to enter from both top and bottom. The boxes are placed one on top of the other in such a manner that air can enter the bottom of each one. This method works out nicely for the small farm. Some more practical arrangement can be worked out, I am sure, for larger farms.

Now let us see what happens to the wheat seed when it gets to the big, modern flour mill. Strong chemicals have to be used to prevent the flour from going bad. Modern wheat does not have the keeping qualities of the old-fashioned kind which was allowed to stand in shocks in the field and mature. Today combines go into the field, cut and thresh the wheat in one operation and rush it direct to the mill. When the wheat stood in the shocks the stem still fed the seed and permitted it to season properly; later it was threshed. Some farmers used to store their wheat unthreshed in the barn and thresh it as needed. The seasoning and maturing process still went on in the barn. This is the method we have restored on our farm. In England, I am informed, seed merchants and maltsters will not handle grain from a combine harvester as the wheat always has to be dried, and it so easily loses its germinating power.

A manager of a big flour mill, after reading an article in *Organic Gardening* concerned with this problem, wrote advising me how true it was and stated that they were at their wit's end trying to figure out methods to give the flour keeping qualities. Their main recourse was to chemicals.

It must be borne in mind that where wheat is grown in a fertile soil enriched only with organic matter, the grain has a far greater keeping quality as compared with the product of a chemicalized soil. Figure it out for yourself—chemicals in the soil, chemical treatment of the seed, chemical sprays on the harvest in the barn, chemicals applied to the wheat at the mill—does this seem like a fit product to go into the human or animal digestive system?

This discussion has not taken into account the strong chemical bleaches used to whiten flour. Labat, a French authority on the subject of bread, says, in the *Bulletin of Hygiene,* "The danger of chronic intoxication following the persistent use of bread made with flour that has been bleached and artificially matured by means of chemical improvers is held to be sufficiently well established to make the absolute prohibition of the use of any chemical improver in France highly desirable."

Now we go a step further. What happens to the flour when it arrives in the bakery? In the book *Eat, Drink and be Wary* by P. J. Schlink, appears this statement:

> "The amount of lead which reaches the consumer's stomach via bakery goods must be enormous and gravely menacing to the health, judging from the foregoing statement, which describes a typical condition; and from the fact recently disclosed by an analysis conducted for Consumers' Research, that ammonium carbonate, used commonly as a leavening (gas-forming) agent in certain baker's goods such as cookies and cakes, contained, as obtained from a commercial bakery in a large Mid-western city, the enormous and threatening proportion of 70 parts per million of lead. The early decline of vitality and the incidence of chronic disease in people who live predominantly on products of commercial bakeries could be explained almost on this type of finding alone. On the other hand, research in such contaminations is made very difficult so that a consumer's organization has the utmost difficulty in even getting samples of the very special and very peculiar materials used in the commercial baking industry—the various gelatinous pie-filling mixtures, the highly colored and synthetic cornstarch custards that go into pies, eclairs and cream puffs, the special low grades of chocolate, the dyestuffs, the egg powders, the

paraffin-like hydrogenated fats especially devised and distributed to commercial bakers and to them alone."

Coming back to the farm, there are dozens of other chemical applications which are part and parcel of the chemical system of farming. Chemicals are used in raising seedlings: to kill off organisms that produce fungus and other diseases. They also kill off beneficial organisms. In order to preserve the wood of boxes in which seedlings are raised, they are often painted with creosote, a highly irritant substance, often deadly to plants. Tests conducted by the Department of Botany at Massachusetts State College with young tomato plants in boxes painted with creosote showed definite injury to the plants. Plants have even been known to be injured by creosote fumes. It is better to paint these boxes occasionally with linseed oil, or not at all.

Countless other instances of dangerous chemicals used on the farm could be given without touching on poison sprays for garden and orchard which are discussed elsewhere.

Before concluding, I want to mention the case of Arthur Yates & Co., Ltd. of Auckland, New Zealand, a large seed and garden supply distributing house. They issue a colorful 65-page catalogue every year and have distributors in every town in New Zealand. Both New Zealand and England are, of course, far ahead of the United States in using composts.

In their 1942 catalogue, a lengthy introduction attacks both the chemical interests and the chemical system of farming:

"In the Waikato district (one of the greatest users of artificial water-soluble stimulants such as Sulphate of Ammonia and Super), the average useful life of a milking cow, as the result of diseases such as contagious abortion, sterility, etc., is now only three to four milking seasons, which is less than half that of the 'good old days' of correct living and manuring.

"It is hard to realize what a comparatively few years it took to poison our soils and foods. We have now become a people with some of the worst teeth in the world; about 90% of our children grow up with bad teeth. We are told by our medical examiners for military service that we are fast deteriorating into a C3 nation [equivalent to our 4F, Ed.], incapable of producing men fit to fight and defend the country without breaking down. For a country with the finest climate in the world, with ample stocks of food, sunshine hours, plenty of clothing, and an outdoor life, this is an absolute disgrace.

"The truth is that the theory of chemical fertilization was founded on a misconception. It was argued that because a plant is formed of chemicals they only need to be applied to the soil in chemical form for the plants to get all they need. But this was an unnatural process. The fact that plants do not naturally absorb these chemicals in their chemical form, but extract them from living organisms, was ignored by the founders of Chemical Agriculture. That the results of forcing unnatural chemical foods on to the plant world would be nearly as fatal as a diet of the chemical constituents of meat and bread in their original form as lumps of iron, etc., would be to man, was overlooked in the early days of chemical manures—about one hundred years ago."

The article winds up with this remarkable paragraph:

"In concluding these remarks on this vital question that badly needs publicity, we would like to say that, although we sell and supply what our customers demand, we have formed our own ideas by generations of experience of the workings of Nature, and we have taken the liberty of expressing some here, although loss of sales of super, insecticides, plants, trees, etc., will result, through a general improvement in the health of plant life.—Yours sincerely, for our country against 'rackets,'" ARTHUR YATES & CO., LTD.

5

CHEMICAL VS. BIOLOGIC CONCEPTS

The chemical fertilizer industry began with Justus von Liebig, the German chemist, who in 1840 established the foundations of the present gigantic trade in farm chemicals. Up to that time the humus theory had been the guiding basis for agriculture. Liebig impressed the scientists and intellectuals, and, after them, the "practical" farmer interested in quick yields, to such an extent that chemical agriculture has held its sway for over 75 years. The basis for his discovery was an experiment in which he burned some plant matter. In the remaining ash he found mainly phosphorus and potassium. This was the beginning of the NPK school of agriculture, and of what Sir Albert Howard calls "the NPK mentality" in agriculture. (N—nitrogen; P—phosphorus; K—potash).

It was announced as an epoch-making discovery and since experiments showed that the application of NPK increased plant growth its use spread like wild-fire. All the farmer needed was a few bags of chemicals and he could get results in a matter of days. Liebig completely overlooked the biologic life of the soil, the bacteria, the fungi, the earthworms and the moulds, because he was not a farmer. He was a chemist. He was evidently not aware that plants do not *naturally* absorb these elements in their pure form but get their food after it has been worked on by living organisms that prefer organic wastes as their raw material. Russell, in *Soil Conditions and Plant Growth,* shows that many of Liebig's ideas were entirely incorrect. The ash does not give a correct measure of what the plant needs. Turnips, for example, seem to need phosphates, yet their ash shows very little of this element.

Big industry was naturally quick to exploit the "miraculous" discovery. It enabled it to turn into quick profits the worthless chemical wastes of the smelting plants and gas works. Experiments soon undertaken by the Rothamsted Agricultural Experiment Station in England, by which wheat was grown year after year on the same plot with chemical fertilizers lent the stamp of official authority to Liebig's theory. (As previously noted, Rothamsted used new seed from outside sources each year, and their experiments would not have had the same results if they had used wheat seed from the crops under experiment; for continued use of artificial fertilizers undoubtedly causes seed to lose its reproductive potency.) The fact that great potash deposits were soon discovered in Germany and elsewhere helped in establishing the chemical fertilizer idea.

Viscount Lymington in his book *Famine in England* shows the extent to which the Germans practiced what they preached:

"Germany . . . in 1914, on poorer average soil and with a worse rain fall, produced more per acre than we did. While this was partly the result of good farming, it was due more especially to her pioneering work as the protagonist of artificial fertilizers. The appalling result of this showed in 1914-18 when a large part of her population starved in spite of her being able on paper to produce 85 per cent of her total food requirements. War cut off or deviated her supplies of mineral fertilizers. In spite of being in 1914 the best-fed nation (including Great Britain) in Europe, she starved because the land without artificial fertilizers failed to keep up her crop production. Her livestock decreased in numbers and there was no reserve of fertility. This could not have happened if she had farmed her land more organically and not over-stimulated it by artificial means. As soon as the pills were cut off constipation of the soil set in. If we were to return to good and only semi-intensive organically manured agriculture we would be saved from that disaster, and we would be independent of artificial aids in future crises."

The fertility of soil involves many factors besides mere nitrogen, phosphorus, potash and calcium. There are physical and biological conditions and factors as yet unknown. The chemical concept for example, completely ignores the mycorrhizal phenomenon as a factor in soil fertility. The great chemical minds of agriculture run aground in their reasoning when they attempt to determine the exact chemical analysis of composts and point out weaknesses in their chemical make-up as compared with chemical fertilizers. *It is not what a compost is but what it does.* This thought is very effectively expressed in Balfour's *The Living Soil:*

"It was found necessary to arrange a special experiment in order to demonstrate that the beneficient effect of compost bears little relation to its chemical composition, and that its biological reactions are of more importance than either its physical properties or the plant nutrients which it may contain."

Here she is referring to the testing by an agricultural chemist of two kinds of compost, one entirely organic, the other made with the addition of chemical activators. The chemists pronounced the chemically activated type as much the best, calling the organic compost "an organic manure—of comparatively little value," yet in an experiment that followed, the organic compost showed the better results, although it lacked the NPK so necessary to the chemist's peace of mind. In my opinion, after this question has been studied many more years, it will be shown that nitrogen, phosphorus and potash *are* necessary, but not in anywhere near the quantities demanded by present-day agricultural chemists.

A reader of *Organic Gardening* asked whether by using certain particular kinds of raw materials in making compost he would end up by having humus rich in potassium, or by varying the raw materials obtain a product favoring nitrogen or phosphorus. Sir Albert Howard answers this as follows:

"Your correspondent is looking at a complex biological problem from the angle of chemistry which deals only with a minor aspect of the question. A small tail is wagging a big dog! The problem of crop production is to feed the soil population—earthworms, fungi, and bacteria and so forth—who in turn feed the plant. If we look after the soil population, everything else will follow. If we try to feed the plant direct, as with chemical fertilizers, Nature's reaction is loss of quality followed by disease. Properly prepared compost, therefore, is all that the soil needs."

According to Howard, "plants seem to need very little phosphorus, if that phosphorus is in the organic condition." Dr. Ehrenfried

Pfeiffer in *Bio-Dynamic Farming and Gardening* comments, "There is, for example, the nitrogen problem. It is plainly evident that there is an antagonism between the nitrogen produced by bacteria and the nitrogen coming from the use of mineral fertilizers. Legumes "forced" with artificially applied nitrogen develop no bacteria. Clover on meadows fertilized with ammonium sulphate disappears."

A study made by Thomas H. Mather of the University of Alberta, Edmonton, shows that where superphosphate was used as the chemical fertilizer in a crop of various kinds of hay, "practically the entire increase in total phosphorus . . . was confined to the inorganic phosphorus fraction." There was no increase in the organic phosphorus of the plant. Experimental work is needed in this field, but there is much evidence available to show the great difference in the form of minerals that come out of a mine and those produced by green manure or composts.

In the Journal of *Industrial Engineering Chemistry,*[1] R. C. Collison in an article entitled "Inorganic Phosphorus in Plant Substances," shows that the seeds of plants contain more phosphorus in organic combination than the other organs of the plant. This is a method nature uses for the perpetuation of the species. The seed must have the most strength, and the most potency in order to be able to carry on into the next generation. And when there is a certain amount of organic mineral available, as much as is possible of it is conserved for the seed. When chemical fertilizers are used there is an insufficient amount of organic mineral produced to give the seed the necessary reproductive strength. As a consequence, seed or plant strains die out.

In April, 1941, an experiment of tremendous significance was described by Professor Glen Wakeham of the University of Colorado before an American Chemical Society meeting at St. Louis. This experiment proved that garden beans and peas grown in poor soil had a greater mineral content than those grown in rich soil. By rich soil was meant soil that was "enriched" by the usual chemical or commercial fertilizers with the possible addition of manure.

Professor Wakeham said: "While millions of dollars are spent annually in research to produce greater crop yields, and other millions to

[1] Vol. 4, 1912.

discover ways of disposing of the surplus crops thus produced, almost no attention has been paid to the nutritional values—particularly with regard to essential minerals—of the artificially forced and 'improved' crops obtained. If the present observations prove to be examples of a more or less general phenomenon, crops grown on 'poor' soils should command higher prices than those grown on 'rich' soils.

" 'More fantastically,' it has been suggested 'that tribes and nations living in agriculturally poor countries—high, windswept plains, mountainous districts and arid semi-deserts—have usually conquered the inhabitants of the rich river valley and coastal plains because they have 'more iron in their blood and more grit in their bones.'

"For years nutritionists have pointed out that many average American diets are likely to be deficient in calcium, phosphorus, iron, iodine, etc., causing such deficiency diseases as rickets, anemia, dental decay, and 'simple goiter.'

"Thousands of analyses have shown that while standard foods will usually contain fairly uniform amounts of the great food elements—proteins, fats, sugars and starch—their mineral contents may vary widely. Some samples of spinach, for example, contain seven times as much iron as other samples."

The question of vitamins in this connection will be discussed later.

Chemical examination of soils does not seem to give a true measure of soil fertility. Two soils may seem to be identical in chemical analysis, yet the yields of their crops differ radically. An interesting situation in Pusa, Bengal, India, showing the fallacy of the chemical concept of agriculture, is described by Howard in the *Testament:*

"The soil is a highly calcareous silt-like loam containing about 75 per cent of fine sand and about 2 per cent of clay. About 98 per cent will pass through a sieve of 80 meshes to the linear inch. There is no line of demarcation between soil and sub-soil; the sub-soil resembles the soil and consists of alternating layers of loam, clay and fine sand down to the sub-soil water, which normally occurs about 20 feet from the surface. The percentage of calcium carbonate is often over 30, while the available phosphate is in the neighborhood of 0.001 per cent. In spite of this low content of phosphate, the tract in which Pusa is situated is highly fertile, maintaining a population of over 1,200 to the square mile and exporting large quantities of seeds, tobacco, cattle, and surplus labour without the aid of any phosphatic manures. The facts relating to agricultural production in this tract flatly contradict one of the theories of agricultural science, namely, the need for phosphatic fertilizers in areas where soil analysis shows a marked deficiency in this element."

The same principle is illustrated by Dr. Pfeiffer:

"Tobacco is rich in potassium when it grows in a soil poor in potassium, and vice versa. The wood and bark of oak trees are especially rich in calcium (up to sixty or more per cent of the ash is CaO). Moreover, they can grow in sand—that is, in a soil poor in calcium—and despite this are able to accumulate calcium. And, their content may reach 60 per cent."

Dr. Pfeiffer says further:

"The author is acquainted with the studies of noted research institutions which report that on large test areas no diminution of phosphoric acid has occurred despite intensive cultivation of the land; although no phosphoric acid was added other than that provided by the normal manuring of the land. It was observed that the P_2O_5 content of the same soil was subject to fluctuations at different times of the year—an indication that one is dealing here with organic processes within the microscopic world of the soil."

In an experiment at the Pennsylvania Agricultural Experiment Station, described in *Bulletin 398*, "Soil Testing," there is illustrated the fact that many chemicals applied to the soil become "fixed"; that is, unavailable to plants. It states:

"Small applications of superphosphate, 325 pounds (to the acre), were completely fixed. Large applications, 625 pounds, were not quite all fixed, while the largest application, 1200 pounds, permitted a level of 50 pounds per acre on June 17, 1937. The plot receiving 10 tons of manure in addition to 625 pounds of superphosphate had the highest level of extractable phosphate, even higher than that receiving 1200 pounds of superphosphate without manure. Noteworthy is the fact that generally when there is a discrepancy between like treated plots, as between tier B and tier D, that one having the highest level is the one highest in organic matter."

Note the effect of organic matter in providing available phosphorus.

One of the members of the staff of *Organic Gardening,* Roger W. Smith, has expressed very succinctly the new idea of making soil tests, not by chemical but by biological means:

"Much has been done—and most of it very unsuccessfully—in the way of attempting to measure the organic content of the soil. Almost all this work has been carried out by chemists using chemicals. These earnest men naturally did what they could to reduce the organic content of the soil to some given chemical 'structural formula.' Their tests, as far as we can find, not only failed, but what is more significant, they failed themselves to understand the real object of their search.

"Their quest differed from ours in that they sought to measure soil organic matter; we seek to measure soil fertility which, as you well know, is a very different thing. They wanted to do it the chemical way; we seek to measure soil fertility by measuring the growth it will produce in a short time. If we can pin this growth down to the development of some given, swift-growing, micro-fungi we believe we will have a really valuable yardstick by which to measure soil fertility."

Dr. Smith of the Florida Agricultural Experiment Station agrees with the opinion just quoted. He sums up the situation as follows:

"Since the soil micro-organisms are plants it seems logical that the biological tests for the availability of plant food constituents and the determination of the fertilizer needs of soil should be more satisfactory than chemical tests. Although several of these tests have been under investigation from time to time, no procedure has been developed which does not leave much to be desired. Soil microbiology is still a virgin field."

This is the kind of experimental work that our agricultural colleges should undertake. Modern agricultural research needs to make an about-face, away from the conveyor-belt type of farming methods, away from intensive, chemical theories, and forward to a biologic conception of soil fertility. Scientists, it seems, must work out a reason or theory for everything that happens. Farming has been too much reduced to a chemical formula by unimaginative, doctrinaire men.

The good farmers of by-gone centuries in every continent were no more aware of *why* humus, crop rotation and compost and manure were valuable in keeping land fertile; they knew it was from use and observation and experiment, just as the Egyptian knows that the silt of the flooding Nile fertilizes his acres.

The agricultural scientist of the future will have to be open-minded, able to close his mind to formulas, to search the clues offered by ancient history, folk-lore, and every suggestion that comes his way, to learn the why and the wherefore of soil fertility and conservation —and their relation to the production of food that is full of life-giving nourishment.

6

YOU HAVE TO KNOW *YOUR CHEMISTRY*

The use of chemical fertilizers on the land is such a complicated procedure that many farmers do not learn to handle them properly. The result is that they follow hit-or-miss methods which are wasteful of time and money and do not attain or even approximate the results claimed by the fertilizer sellers. The average farmer doesn't pretend to understand the chemical requirements of his farm and since he can't have the county agent at his elbow all the time, he uses his own judgment and is apt to suffer by it.

Some farmers think that if a small amount of chemical fertilizer is good, a lot is better. This brings disaster in many cases. An acquaintance of mine bought a popular brand of chemical fertilizer for his flower garden. He carefully followed directions as to the proper quantity to use. When he had finished, he found he had used only half the bag. He decided to use up the remainder in order to get it out of the way. The result was that he burned up many of his best plants. In *North-West Gardens,*[1] there is a warning:

[1] June, 1944.

"Because commercial fertilizers are so highly concentrated, they must not be allowed to come in contact with the plant, so the trench is dug from two to four inches away from the stems. Also, care must be taken to remove immediately any that may be spilled on the leaves, for if allowed to remain, burning may result. Many gardeners do not realize the potency of commercial fertilizers and use far too much. This is both dangerous and wasteful; dangerous because roots can be burned by large amounts not properly blended with the soil, and wasteful because plants can only take up so much food at a time, and when too heavy an application of fertilizer is made, a large portion leaches away before the plants can use it. It is far better to apply small amounts more often."

Special attention should be given to the "double-strength" fertilizers now in vogue. Put up this way to save freight, they pose unusual problems for the unwary user. An extremely common fertilizer, which requires a great deal of expertness in using, is nitrate of soda. H. R. Ely in *The Practical Flower Garden* [2] says:

"Then there is nitrate of soda, which does not stimulate root-growth, but is valuable in producing rapid increase in bloom and tends to give more brilliant coloring to the blossoms. This product should be regarded, however, as a quickening tonic, for use somewhat as nitroglycerine is prescribed by the medical profession.

"No fertilizer will produce such quick results as nitrate of soda if properly used, but, if used too freely, probably no other fertilizer can damage the plants so quickly."

A very difficult matter is the test to decide how much NPK is required. The reason is that not only are there seasonal variations but variations in different parts of the same field. Phosphorus has been shown to test lower in the summer than in the fall or winter. Because of the fact that much time cannot be given to each test and because of technological difficulties many errors creep in.

Nature is an experienced soil chemist, as well as soil biologist. She works deftly and harmoniously. She mixes her brews with an experience stemming from a background of millions of years. Say what you will, man rarely figures out his discoveries from a preconceived theoretical plan. They result either by accident or by trial-and-error. Man

[2] The Macmillan Co.

throws an excess of potassium in the soil and is not aware that this precipitates the soil magnesium. Magnesium is badly needed, a lack of it causes diseases in rye, oats, tobacco and potatoes. In Alabama it was found that on certain clay soils the application of phosphates caused iron chlorosis. Man sprays the soil and unknowingly there accumulate in that soil, arsenic, copper, lead and zinc, which after awhile prevent even cover crops from growing. Trial-and-error, but mostly error.

Experiments show that young seed cannot tolerate too much chemical fertilizer. After the seed has grown somewhat it can tolerate a higher concentration. Methods have been evolved where machines apply the fertilizers in bands away from the seeds where the plant can get at it later, but results are at best erratic. Trouble is encountered also over accumulated residues from previous chemical applications which affect plants at the seedling stage. Many old citrus orchards in California have large accumulations of phosphates and potash. This causes trees to become infected with a disease called brown-rot gummosis.

An important experiment performed by Merkle and Dunkle at Pennsylvania State College reported in the *Journal of the American Society of Agronomy* (1943) revealed that in more than 20% of the commercial greenhouses encountered by the experimenters, there were excessive amounts of residual fertilizer salts in the soil. They say, "From these considerations it is obvious that many of the greenhouse soils examined had salt concentrations entirely too high to be satisfactory culture media for the plants grown, and that any drop in the moisture content of the soil or unusual transpiration demand might react unfavorably upon the plants." What chance has the small gardener to sail safely amidst these rocks and shoals when experienced men cannot do so?

A reader asks his gardening magazine for advice. His tomato plants grow tall without ripening fruit. He is told that his soil contains too much nitrogen and not enough phosphorus. So he will go hard on the phosphorus. Next year his plants will suffer from symptoms of an oversupply of phosphorus. The fact is that there is no single formula for tomatoes. In the April, 1944, issue of *Farm Research,* a Cornell University publication, one finds:

"In fertilizer experiments with cannery tomatoes it has been demonstrated that no one formula can be recommended for all soils. The highest yields were obtained from the highest nitrogen ratio on Ontario loam at Geneva, from the highest amount of phosphorus on Fulton silty clay at Fredonia, and from the heaviest potash application on Palmyra gravelly loam at Marion."

The question as to the correct amount of chemical fertilizer to use is an involved one. Statistics show that about three times as much nitrogen is being used today as before the war. Yet yields have not increased. In 1942 the general artificial fertilizer consumption was twice the quantity produced ten years before with no crop increase to show for it.

In *Science and Appliance* there is some remarkable comment on a pamphlet published by Ohio State University, in 1936:

"This pamphlet, entitled *Our Heritage, the Soil,* was written by three members of the department of agronomy at the Ohio State University. In a careful, mathematical fashion, the authors considered the productivity of the soil as indicated by agricultural statistics of Ohio over a period of sixty years. They came to the shocking conclusion that, although superficially we appeared to be reaping larger returns per acre at the end of the period than at the beginning, the increase in productivity had not been nearly so great as might have been expected. We were farming better, not getting a great deal more. To make the matter appear much more alarming, in the interval some of the poorest lands had been retired.

"The situation, as stated simply, is serious. The permanent value of the soil appears to have been impaired. The land may be stimulated to produce fairly good yields, but the stimulation is only temporary. The capital itself has deteriorated.

"It is not on fertility alone that the soil has suffered. Its tilth or working character has suffered. Lands have been reported as much heavier and more compact than formerly, harder to drain and more liable to bake to bricklike solidity.

"The situation is summed up in this fashion: The natural productive capacity of the land has been deteriorating at a rate almost fast enough to offset all of the improvements in soil and crop management."

The state of Iowa, which has not yet made as much "progress" in destroying the soil's fertility by the use of chemicals as most other states, used only 17,000 tons of fertilizers in 1941. New Jersey in that year used 184,000 tons. When it is realized that the farm acreage of New Jersey is only one-eighteenth that of Iowa it can be seen that they are using 200 times more fertilizers than Iowa. To a great extent this is due to the intensive type of truck-farming of New Jersey which furnishes vegetables to New York City and Philadelphia with the use of prodigious amounts of chemicals to get early, premium-enjoying maturity (at the expense of the big-city dwellers).

One is afraid to think of what is going to happen after World War II. After the first World War the factories that were built to fix nitrogen from the air to make explosives had to find a use for the product of this equipment. This was accomplished by big-scale advertising programs. In this war the capacity to produce cheap ammonia is being greatly increased on account of war needs. In fact this situation is being recognized by Government scientists who are now developing methods for post-war use, to see how more ammonia can be used by the average farmer.

The chemical companies in recent years have been advocating the use of humus along with artificial fertilizers. Typical of such items is the one in *The American Fertilizer* of April 8, 1944, called "Humus 'Carries the Ball' for Fertilizer." In this article George Serviss says:

"Chemicals alone, though, are not enough for a successful, permanent agriculture. Attention must be given to other things and one of the most important is organic matter. Failure to replenish organic matter as a result of economic pressure on one hand and too great faith in chemical fertilizers as 'cure alls' on the other, has already resulted in declining yields on some soils. This has provided some 'thunder' for an organic matter 'cult,' the members of which would dispense with the use of chemical fertilizers and rely entirely on organic residues and manures for maintaining soil fertility. This, any one at all versed in Soil Science will tell you, would have serious consequences for a consuming population of 140 million people a few years hence, even if farmers' pocket-books could stand it."

By Mr. Serviss's definition, I belong to that "cult"—membership in which has been held by good practical farmers from Adam's time on down the centuries, so, if by "cult" a small, esoteric society is meant, then I wager that time will rather call the followers of the chemist Liebig the cultists.

Howard has given thought to the matter of using chemicals and humus in combination. In *Country Life,* he stated:

"Can humus supply everything the crop needs? Is there any advantage in supplementing humus by artificials? The answer to the first question is clear and definite. In all cases where the soil has received sufficient humus to make it really fertile there is no appreciable response to artificial manures. The maximum yield can be obtained without chemicals. This was one of the interesting results obtained by Mr. A. J. Hosier after he had got the worn-out downland of Wiltshire back into condition by his outdoor method of milk production. When he began operations in 1924 he could write his name on the fields with chemicals. Once they became fertile there was no response even when a complete fertilizer was applied.

"Before the condition of real fertility has been reached and insufficient humus or imperfectly prepared humus has been used there is no question that the addition of chemicals does increase the yield. As this state of affairs applies to very large areas in this country it must be conceded that there is a temporary use for artificials. They will enable crop-production to be maintained during the period while the supplies of humus are being increased. While this is in progress a progressive reduction in the amount of artificials used can be made.

"If artificials can supplement humus why not be content and make use of such substitutes? The answer to this question is provided by the important principle that what matters most in crop production is the synthesis of protein in the green leaf. When this takes place by means of humus everything goes well—disease resistance, high quality and the power to reproduce the species all follow as a matter of course. Where, on the other hand, a substitute phase in the shape of artificials is interposed in the formation of protein, trouble ensues—the power to resist disease, high quality in the produce as well as the stability of the species are lost. Crop production under these conditions fails."

There is another confirmation of this principle in the *Farmers' Weekly* of April 7, 1943, published at Bloemfontein, South Africa. The item is by N. W. Ayson:

"I have read Mr. Viedge's very sensible letter in your issues of March 17, and can tell him that I have been pouring in stable manure and rotten straw into my garden for the last 20 years. At the beginning my garden was just pure yellow stiff clay (just like yellow soap); now it is almost black heavy loam and does not require any more manure for some years. Thinking to get even better crops I have added to beans, mealies, beet and so on, artificial fertilizer containing phosphates, but this had no effect whatsoever as against those parts not so treated."

There is today a strong back-to-the-land movement. People are going on farms who know absolutely nothing about agriculture. How simple it would be if they farmed by the organic method. They would require very few tests, no complicated chemistry, no scientific training; merely the simple process of loading organic matter, humus, into the soil as in a savings bank, increasing their balance from year to year.

PART FOUR

Health or Disease from Food

I

HUMAN HEALTH AND COMPOSTS —THE "MEDICAL TESTAMENT"

My interest in the work of Sir Albert Howard began when I read several years ago, in an English health magazine, the details of an experiment in the feeding of boys at a large preparatory school near London. These boys were fed on vegetables grown by the Howard method, with humus only and the experiment was "accompanied by results," says Howard in the *Testament* "of considerable interest to parents and to the medical profession. Formerly, in the days when artificials were used, cases of colds, measles, and scarlet fever used to run through the school. Now they tend to be confined to the single case imported from outside. Further, the taste and quality of the vegetables have definitely improved since they were raised with humus."

The evidence of these results is contained in a letter written to Howard by the headmaster of the school.

This item spurred me on to get in touch with Howard, to read and study his classic book, and finally to purchase the farm on which we now live. Since then many cases have come to my attention which satisfy me that the use of chemical fertilizers in growing our food is a major cause of the present lack of health in human beings. A few years ago *The Lancet,* English medical journal, reported a case in New Zealand similar to that of the English school: "In 1936, Dr. G. B. Chapman, of the Physical and Mental Welfare Society of New Zealand, persuaded the authorities of a boys' school hostel to grow their fruit and vegetables on soils treated with humus. This has since been done, and a striking improvement is reported in general health and physique, particularly as regards freedom from infections, alimentary upsets and dental caries."

The N. Y. Times on June 30, 1940, also discussed this case, identifying it as the Mount Albert Grammar School. According to the

Times: "Dr. Chapman advised that a change should be made from vegetables and fruits grown in soil fertilized by chemicals, to produce raised on soil treated only with humus. The results were startling. Catarrh, colds and influenza were greatly reduced and in the 1938 epidemic of measles, the boys had only mild attacks whereas new admissions succumbed readily."

Still another experiment along the same lines was conducted at the College of St. Columba, Dublin, Ireland. An account of it is given in a letter written to the London *Times* (Feb. 28, 1944), signed jointly by the Archbishop of Dublin; Lord Iveagh; E. H. Alton; Roland L'E. Bryce; C. M. Dobbs and Hugh A. C. Maude, in which they say:

"In the reorganization of education we humbly submit that some provision should be made for the practical teaching of agriculture, arboriculture and horticulture on a wide scale in many of the schools, which have the good fortune to be situated in country districts, so that boys and girls, who will not necessarily become farmers, may have some knowledge and appreciation of this all-important problem. Agricultural education should not be left to the specialist students in colleges given over to that purpose, but should have a far wider circulation.

"May we quote from our own experience as Fellows of the College of St. Columba? This college, one of the smallest public schools, recently acquired some long-neglected land, and with its own small property established a profitable mixed farm. It is worked by a committee of masters, boys and workmen. Much of the work is done by the boys, working with carefully selected men, none of it during class hours. The greater part of it is practical manual work in the open air. Once a week boys voluntarily attend agricultural lectures in the biology laboratory.

"All farm, garden and house waste of an organic nature is turned into compost. Over 100 tons was made last year. The soil is continuously being enriched, while drainage, weeding, animal husbandry, and tree planting are methodically carried out. Not only is the school provided with fresh vegetables, milk, beef, mutton, and wheat, but the general health is noticeably better with a greater immunity from infectious disease. Also, the standard of work, judged by public examination, is improved. We should mention that all surplus farm produce in the holiday periods finds a ready local market. The College of St. Columba has no reason to regret the experiment. For these reasons we would submit this information to your valued paper."

Another significant case is described in *The Cross and the Plough,* a magazine published by the Catholic Land Federation of England and Wales:

"At the Winsford Bacon Factory in Cheshire, one of the most efficient organizations in the kingdom, the Co-operative Wholesale Society sometime ago decided to try out the principles underlying *The Medical Testament* of the Cheshire doctors and to provide the staff, some 150 in number, with two good meals a day at cost price, the fresh food used being raised on fertile soil. For this purpose an area of land round the factory was brought into a high condition of fertility by the help of humus made from the wastes of the factory and of the land.

"Things were going well, the manager was enthusiastically devoting to this work his great powers of organization and initiative, the local doctors, as well as the local notables, were watching the developments with interest and sympathy. I sampled one of these meals, which could not be bettered anywhere. At Winsford I felt a fire was being lighted which would soon spread throughout the length and breadth of the land for the reason that enlightened industry has for some time been trying to find some means by which cordial working arrangements can be established between the management on one side and the labor force on the other. One direction is obviously for capital to use its powers to improve the health, well-being and contentment of the workers. Winsford was providing such an example of constructive social service.

"Then the blow fell. The Ministry of Food decided to close this factory under the scheme for the concentration of industry, quite oblivious of the fact that important developments were in progress in the processing of pigs, in the preparation of various animal extracts needed in medicine, as well as the provision of ideal meals for the workers already described. Naturally protests were raised.

"The Society's appeals to the Tribunal and to the Ministry of Food were disallowed. The Cheshire Panel Doctors also took up the case and laid the facts before Lord Woolton. In the correspondence which followed the Ministry cut a very sorry figure and were literally reduced to mincemeat. Had it been possible to take this case to the High Court, His Majesty's Judges would have castigated the bureaucrats in no uncertain fashion. The decision to close the factory was obviously made without a knowledge of the facts—in other words, on a fragment only of the case."[1]

[1] Sir Albert Howard reports to me that the Winsford factory has since been reopened

Another experiment worthwhile recording is that of Dr. J. W. Scharff, when he was Chief Health Officer at Singapore. The facts are contained in a letter he wrote in September, 1942, to the *News Letter on Compost* published at Holmes Chapel, Cheshire, England:

"From January, 1940, until January, 1942, I had a unique opportunity, due to war-time needs, of watching the progress of a campaign for growing vegetables and seeing that they were eaten by a labor force of nearly 500 Tamil coolies. These men were employed by the Singapore Health Department in various parts of the island of Singapore. As soon as England became involved in war it became possible to allocate an area totalling in all about 40 acres of vegetable allotments on favorable terms to the men engaged on sanitary duties. My laborers were granted these allotments on condition that they prepare compost and used the vegetables and fruit grown therein for themselves and their families only. Sale of the produce was not allowed. Thus it was ensured that these goods were used at home. The local Agricultural Department lent their inspectors and staff to teach the men how best to grow vegetables, and demonstrations in cooking and preparation of the foodstuffs were organized for each of the labor settlements. Compost making was started on a large scale and during the months previous to the opening of the campaign a supply of over a thousand tons of compost was ready to launch this great experiment.

"During the course of the ensuing months, apathy and indifference on the part of the laborers, gave way to interest and enthusiasm as soon as it became apparent how well plants would grow on soil rendered fertile with compost. A number of vegetable shows were arranged, at which the healthy produce of fertile soil was exhibited and prizes were awarded. Within six months the accumulated stock of compost was used up and more active steps were taken to augment the supply as well as to satisfy the growing demands of other enthusiastic gardeners inspired by the achievements of my men.

"At the end of the first year it was obvious that the most potent stimulus to this endeavor was the surprising improvement in stamina and health acquired by those taking part in this cultivation. Debility and sickness had been swept away and my men were capable of, and gladly responded to, the heavier work demanded by the increasing stress of war. But for the onslaught of the Japanese which overwhelmed Malaya I should have been able to present a statistical record of the

and the canteen meals are being continued: "The results are startling; the general health of the worker has improved; there is less absenteeism; the esprit de corps is excellent; work is going with a swing; the workers buy all the bread and other food left over."

benefit resulting from this widespread effort of vegetable culture on compost such as would astonish the scientific world. The results were all the more dramatic in that I had not expected this achievement.

"The numbers taking part in this venture were so large as to preclude any possibility of mistake. It might be argued that the improvement in stamina and health amongst my employees was due to the good effect of unaccustomed exercise or in the increased amount of vegetables consumed. Neither of these explanations would suffice to explain the health benefit amongst the women, children and dependents of my laborers who shared in this remarkable improvement. Shortly before the tragic disaster which has brought Singapore within the hateful grasp of the Japanese invader it became apparent that the health of men, women and children who had been served consistently with healthy food grown on fertile soil was outstandingly better than it was amongst those similarly placed but not enjoying the benefits of such health yielding produce. An oasis of good health had become established, founded upon a diet of compost-grown food.

"This has served me as an inspiration to carry on with this work in whatever part of the world it may now fall to my lot to serve mankind."

In *The Living Soil* Lady Balfour relates her own experience. She used to have rheumatism and frequent colds. "After changing to compost grown food my rheumatism disappeared, and I now practically never have a cold. I have now been over a year without even a slight one."

Probably the one event that most focussed public attention to the whole problem of compost vs. chemical farming was the dramatic stand taken by the 600 panel doctors of Cheshire in issuing their now world famous *Medical Testament*. Parts of it have been quoted in American journals, but it is too valuable not to be fully accessible, and because of its keystone importance, I am reprinting it here in full:

THE "MEDICAL TESTAMENT"

After more than a quarter of a century of Medical Benefit under the National Health Insurance Act, we the Local Medical and Panel Committee of Cheshire, feel that we are in a position to review our experience of the system.

Constituted by the statute to represent the panel of an area, such a committee is in touch with all the family doctors—in the case of Cheshire some 600—within and on its borders.

How far has the Act fulfilled the object announced in its title—"The Prevention and Cure of Sickness"? Of the second item we can speak with confidence. If "postponement of the event of Death" be evidence of cure, that object has been achieved: the greater expectation of life which is shown by the figures of the Registrar General is attributable to several factors; but certainly not least to the services of the panel. The fall in fatality is all the more notable in view of the rise in sickness. Year by year doctors have been consulted by their patients more and more often, and the claims on the benefit funds of Societies have trended to rise.

Of the first item, "The Prevention . . . of Sickness" it is not possible to say that the promise of the Bill has been fulfilled.

Though to the sick man the doctor may point out the causes of his sickness, his present necessity is paramount and the moment is seldom opportune, even if not altogether too late for any essay in preventive medicine. On that first and major count the Act has done nothing.

We feel that the fact should be faced.

Our daily work brings us repeatedly to the same point: "this illness results from a life-time of wrong nutrition!"

The wrong nutrition begins before life begins. "Unfit to be a mother"—from under-nutrition or nutritional anaemia—is an occasional verdict upon a maternal death. For one such fatal case there are hundreds of less severity where the frail mothers and sickly infants survive. The reproach of the bad teeth of English children is an old story. In 1936 out of 3,463,948 school children examined 2,425,299 needed dental treatment. Seeing that the permanent teeth develop from the 17th week of pregnancy and that certain foods, accurately known since 1918, are the condition of the proper growth, that is a reproach which should be removed. With it would go the varied host of maladies that spring from diseased teeth. That its removal is practicable is shown by Tristan da Cunha. Most of the population of the little island, people of our race, living on the product of sea and soil, have perfect teeth which last them their lives.

Rickets, for which England was a byword when Glisson described it in 1650, is still with us. Gross deformities are rarer, but the big heads, tumid abdomens, flaccid skins, bulged joints and pinched chests are a commonplace of infancy; and even at school age 3,457 cases of rickets with 6,415 others of spinal curvature were found in 1936 by the School Medical Officers in 1,727,031 inspections.

Yet its prevention by right feeding is so easy that every dog breeder knows the means.

Rickets is a heavy contributor to the C3 population. The Maternal Mortality Committee found that there is much less in Holland where butter, milk, and cheese are plentiful and the women by virtue of their generally healthy skeletal development are protected against the risks that are commonly faced by women in the industrial areas of England.

Nutritional anaemia is of two kinds, one subtle and apt to happen during pregnancy, the other simple and due to too little iron in the food. It is known that anaemia especially of the latter kind is common, especially among children, and women, who need much more iron in their food than men. An inquiry into the food of 1,152 families showed that 10 per cent spent 4s. a week per head on food, 10 per cent over 14s. whilst four more groups, of 20 per cent each, spent 6s., 8s., 10s., and 12s. respectively. The food of the three lower groups are definitely deficient in iron. It is certain from this that nutritional anaemia amongst the poorer classes is far commoner than is recognized. Here is an example: The blood colour was tested in two groups of school-children, one a "routine sample" of children, the other specially selected on account of poverty. Only half the poor children had a blood colour of 70 per cent of normal. The final item of our indictment is constipation. Advertised aperients are a measure of its prevalence and the host of digestive disorders which result from it are a substantial proportion of the conditions for which our aid, as doctors, is sought. Yet the cause in every case—apart from rare abnormalities—is the ill choice or ill preparation of food. It is true that we are consulted on these conditions when they are established and have to deal with the effects—gall stones, appendicitis, gastic ulcer, duodenal ulcer, colitis, and diverticulitis—of years in which the body has been denied its due of *this* constituent of food or burdened with an excess of *that*. Other means of cure than proper feeding are called for at this late stage; but the primary cause none the less was wrong nutrition.

Those four items, bad teeth, rickets, anaemia and constipation will serve as the heads of our indictment; but in truth they are only a fragment of the whole body of knowledge on food deficiencies which different investigators from Lind and Captain Cook to Hopkins and the Mellanbys have unlocked.

But it seems to us that the master key which admits to the practical application of this knowledge as a whole has been supplied by Sir Robert McCarrison. His experiments afford convincing proof of the effects of food and guidance in the application of the knowledge acquired.

In describing his experiments, which were made in India, he mentions first the many different races of which the population, 380 million, is composed.

"Each race has its own national diet. Now the most striking thing about these races is the way in which their physique differs. Some are of splendid physique, some are of poor physique, and some are of middling physique. Why is there this difference between them? There are, of course, a number of possible causes: heredity, climate, peculiar religious and other customs and endemic diseases. But in studying the matter it became evident that these were not principal causes. The principal cause appeared to be food. For instance, there were races of which different sections came under all these influences but whose food differed. Their physique differed and the only thing that could have caused it to differ appeared to be food. The question then was how to prove that the difference in physique of different Indian races was due to food. In order to answer it I carried out an experiment on white rats to see what effect the diets of these different races would have upon them when all other things necessary for their proper nutrition were provided. The reasons for using rats in experiments of this kind are that they eat anything a man eats, they are easy to keep clean, they can be used in large numbers, their cages can be put out in the sun, the round of chemical changes on which their nutrition depends is similar to that in man, and, a year in the life of a rat is equivalent to about twenty-five years in the life of a human being. So that by using rats one gets results in a few months which it would take years to get in man. What I found in this experiment was that when young, growing rats of healthy stock were fed on diets similar to those of people whose physique was good, the physique and health of the rats were good; when they were fed on the diet similar to those of people whose physique was bad, the physique and health of the rats were bad; and when they were fed on diets similar to those of people whose physique was middling, the physique and health of the rats were middling. . . .

"Good or bad physique as the case might be was, therefore, due to good or bad diet, all other things being equal. Further, the best diet was one used by certain hardy, agile, vigorous and healthy races of Northern India. (Note: the Hunza, Sikh and Pathan.) It was composed of freshly ground whole wheat flour made into cakes of unleavened bread, milk, and the products of milk (butter, curds, buttermilk), pulses (peas, beans, lentils), fresh green leaf vegetables, root vegetables (potatoes, carrots), and fruit, with meat occasionally. Now in my laboratory I kept a stock of several hundred rats for breeding purposes. They lived under perfect conditions; cleanliness, roomy cages, good bedding, abundant fresh water, fresh air and sunlight—all these things they had; and, they were fed on a diet similar to that of a race whose physique was very good. They were kept in stock from birth up to the age of two years—a period equivalent to the first fifty years in the life of human beings. During this period no case of illness occurred amongst them, no death from natural causes, no maternal mortality, no infantile mortality except for an occasional accidental death. In this sheltered stock good health was secured and disease prevented by the combination of six things; fresh air, pure water, cleanliness, sunlight, comfort and good food. Human beings cannot, of course, be so sheltered as these rats were, but the experiment shows how important these things are in maintaining health.

"The next step was to find out how much of this remarkably good health, and freedom from disease, was due to the good food: food consisting of whole wheat flour cakes, butter, milk, fresh green vegetables, sprouted pulses, carrots and occasionally meat with bone to keep the teeth in order. So I cut out the milk and milk products from their diet or reduced them to a minimum, as well as reducing the consumption of fresh vegetable foods while leaving all other conditions the same. What was the result? Lung diseases, stomach diseases, bowel diseases, kidney and bladder diseases made their appearance. It was apparent, therefore, that the good health depended on the good diet more than on anything else and that the diet was only health-promoting so long as it was consumed in its entirety, so long, in fact, as it contained enough milk, butter and fresh vegetables.

"Many more experiments were done which showed that when rats or other animals were fed on improperly constituted diets, such as are habitually used by some human beings, they developed many of the diseases from which these human beings tend to suffer: Diseases of the bony framework of the body, of the skin covering it and of the membranes lining its cavities and passages; diseases of the glands whose products control its growth, regulate its processes and enable it to reproduce itself; disease of those highly specialized mechanisms—the gastro-intestinal tract and lungs—designed for its nourishment; diseases of the nerves. All these were produced in animals under experimental conditions by feeding them on faulty human diets. Here is an example of such an experiment: Two groups of young rats, of the same age, were confined in two large cages of the same size. Everything was the same for each group except food. One group was fed on a good diet, similar to that of a Northern Indian race whose physique and health were good, and of which the composition is given above. The other was fed on a diet in common use by many people in this country; a diet consisting of white bread and margarine, tinned meat, vegetables boiled with soda, cheap tinned jam, tea, sugar and a little milk: a diet which does not contain enough milk, milk products, green leaf vegetables and whole-meal bread for proper nutrition. This is what happened. The rats fed on the good diet grew well, there was little disease amongst them and they lived happily together. Those fed on the bad diet did not grew well, many became ill and they lived unhappily together; so much so that by the sixteenth day of the experiment the stronger ones amongst them began to kill and eat the weaker, so that I had to separate them. The diseases from which they suffered were of three chief kinds: diseases of the lungs, diseases of the stomach and intestines, and diseases of the nerves; diseases from which one in every three sick persons, among the insured classes, in England and Wales, suffer."

These researches were minutely made on a large scale and, but for the food, the conditions of each group were identical and ideal. Their results to our minds carry complete conviction—especially as those of us who have been able to profit by their lesson have been amazed at the benefit conferred upon patients who have adopted the revised dietary to which that lesson points. It is far from the purpose of this statement to advocate a particular diet. The Eskimos, on flesh, liver, blubber and fish, the Hunza or Sikh, on wheaten chappattis, fruit, milk, sprouted legumes and a little meat; the islander of Tristan on his potatoes, seabirds' eggs, fish and cabbage, are equally healthy and free from disease.

But there is some principle or quality in these diets which is absent from, or deficient in, the food of our people today. Our purpose is to point to this fact and to suggest the necessity of remedying the defect. To descry some factors common to all these diets is difficult and an attempt to do so may be misleading since knowledge of what those factors are is still far from complete; but this at least may be said, that the food is, for the most part, fresh from its source, little altered by preparation and complete; and that, in the case of those based on agriculture, the natural cycle

Animal &
Vegetable→Soil→Plant→Food→Animal→Man→
waste

is complete.

No chemical or substitution stage intervenes.

Sir Albert Howard's work on the nutrition of plants, initiated at Indore and carried from India to many parts of the world seems to constitute a natural link in this cycle.

He has shown that the ancient Chinese method of returning to the soil, after treatment, the whole of the animal and vegetable refuse which is produced in the activities of a community results in the health and productivity of crops and of the animals and men who feed thereon. . . .

Though we bear no direct responsibility for such problems, yet the better manuring of the home land so as to bring an ample succession of fresh food crops to the tables of our people, the arrest of the present exhaustion of the soil and the restoration and permanent maintenance of its fertility concern us very closely. For nutrition and the quality of food are the paramount factors in fitness. No health campaign can succeed unless the materials of which the bodies are built are sound. At present they are not.

Probably half our work is wasted, since our patients are so fed from the cradle, indeed before the cradle, that they are certain contributions to a C3 nation. Even our country people share the white bread, tinned salmon, dried milk regime. Against this the efforts of the doctor resemble those of Sisyphus.

This is our medical testament, given to all whom it may concern—and whom does it not concern?

We are not specialists, nor scientists, nor agriculturists. We represent the family doctors of a great county, the county, said Michael Drayton, of "such as soundly feed"; a county which gives its name to a cheese than which there is none better, though to most Englishmen, alas, only a name; a county where the best farming is still possible, which should minister to the needs of its own industrial areas and of a far wider circle. We cannot do more than point to the means of health. Their production and supply is not our function. We are called upon to cure sickness. We conceive it to be our duty in the present state of knowledge to point out that much, perhaps most, of this sickness is preventable and would be prevented by the right feeding of our people. We consider this opinion so important that this document is drawn up in an endeavour to express it and to make it public.

(Signed by the Members of the Local Medical and Panel Committees).

John Kerr (Chairman) — J. Barry Bennett (Hon. Treasurer)
N. A. Boswell (Vice-Chairman) — Lionel Jas. Picton (Hon. Secretary)

This *Medical Testament* may well come to be recognized as the Magna Carta of a new freedom—freedom from the tyranny of illness. It seems to me that no more interesting or more challenging report has ever been issued by the medical profession.

2

IS OUR HEALTH RELATED TO THE SOIL?

America has more hospitals, is served by more physicians per capita, has the highest medical bills, eats more vitamins, uses more synthetic drugs, has the finest medical scientists working in the most elaborately equipped laboratories endowed by great wealth, in the most magnificent educational and research centers found anywhere, and yet—why have draft boards found more ill health in the present war than was discovered during the first World War? Why the pathetic state of under-nourishment found among our school children? More than 40 per cent of the Nation's potential military man power has been rejected for physical unfitness. 9,000,000 men of military age unfit to serve!

A past-president of the American Dental Association, Dr. Arthur Hastings Merritt, recently stated that tooth decay, today, even though a wonderful system of dental care has been worked out, is as bad as it was 100 years ago. This in spite of so-called correct diets, with their generous inclusion of orange juice and other so-called vitamin-rich foods. Nearly every American (95%) needs some dental care; 30% need it badly.

Take the Canadian situation. The second issue of *The Land* for 1944 comments:

"Dr. Albrecht's paper on Calcium, in the previous issue of *The Land* was of amazing interest to Canadians. A recent survey by the Canadian Council of Nutrition shows that 60 per cent of Canadians suffered an 80 per cent calcium deficiency. In this list the deficiency of calcium came second; vitamin B being 90 per cent. Proteins showed a deficiency of 72 per cent. From the content of Dr. Albrecht's discussion the high deficiencies of both calcium and the proteins suggest a more than ordinary significance."

The general public has been lulled into a false sense of security by longevity statistics. According to these figures the average length of life in 1942 was about 64 years. In 1900 it was but 50 years. Thus it would appear that our health is improving remarkably. But when the situation is carefully studied it will be seen that these figures do not tell the whole truth.

Louis I. Dublin, head statistician for the Metropolitan Life Insurance Co., in an article in the *Reader's Digest* says that in 1850 the average length of life was 40 years:

> "Then came the modern sanitary era. Advances in public health curbed the terrible inroads of cholera, diphtheria, tuberculosis and typhoid, and set up systems of protection over water and milk supplies. As a result, the expectation of life in the United States by 1900 had advanced to about 50 years. The 14 years added since that date are the result of an intensive application of our knowledge of disease prevention, and a general and profound improvement in the standard of living."

There are many other factors, in addition to those mentioned by Mr. Dublin, for the present higher longevity figures. Chief among them is infant mortality, which has been tremendously reduced in the past 20 years (from 86 to 40 per 1000). We have better plumbing and sanitation and a water supply system that assures pure water in most homes. No more plagues periodically sweep the nation to bring the longevity figures down. Because of social security measures no one need starve to death today.

We have better surgery procedures in hospitals. To go to a hospital for an operation 50 years ago was almost a sure invitation to the graveyard. Those were the days of the crudest forms of surgery and bloodletting without antiseptics. The public today is being educated in preventive medicine. It is taught not to take a laxative when there are pains suspiciously indicative of appendix trouble, with fewer fatalities from ruptured appendices as a result. Bear in mind too that with increasing skill doctors have learned to keep unhealthy people such as diabetics alive.

But let us break down the longevity situation into its various elements. Such diseases as smallpox, diphtheria, yellow fever and typhoid have been largely eliminated. Deaths from pneumonia and other "coccus" diseases have been greatly reduced by the sulfa drugs and by penicillin. But such degenerative diseases as cancer, nervous disorders and heart ailments are increasing alarmingly. Heart disease has increased about 60% and cancer 90% in the last forty years. Diseases such as infantile paralysis seem to be increasing from year to year and each year seems to bring new "diseases" never heard of before. In February 1944, a Gallup survey showed that 23,000,000 persons had colds.

Dr. Alexis Carrel recognized the problem when he said:

"All diseases of bacterial origin have decreased in a striking manner. . . . But we still must die in a much larger proportion from degenerative diseases. In spite of the triumphs of medical science, the problem of disease is far from solved. Modern man is delicate, 1,100,000 persons have to attend the medical needs of 120,000,000 other persons. Every year, of this population in the United States, there are about 100,000,000 illnesses, serious or slight. In the hospitals, 700,000 beds are occupied every day of the year. . . . Medical care under all its forms, costs about $3,500,000,000 yearly. . . . The organism seems to have become more susceptible to degenerative diseases."

An important statement about the general health of the country was editorialized in *Dental Items of Interest* (1942) by its editor, Paul H. Belding, D.D.S.:

"The state of health of the American people is something of which we should all be ashamed, and to say that this or that condition is responsible for so many rejections falls in the same category as the blind man's description of the elephant.

"If we look at the health problem of the nation from its broad aspects, we find that every day one in twenty of our people is too sick to perform his regular occupation. The average citizen suffers ten days of incapacity annually. The average child is sick in bed seven days of the year, the average oldster, 35.

"Every day 6,000,000 individuals are sick and 42% of them are suffering from so-called degenerative diseases. Degenerative diseases include the chronic, slow-working diseases, the exact cause of which is unknown at the present time, such as hypertension, coronary disease, rheumatism, appendicitis, gall stones, cancer, peptic ulcers, acne, dia betes, nervous diseases, and last but not least, dental disease.

"The evidence is conclusive, that as a nation we are on the down-hill trail that leads to an inferior race. This information is not new to those who have seriously interested themselves in the problem."

General health statistics are in many ways more indicative than tables of longevity. In 1938 Lord Lymington wrote, "workers insured under the National Health Insurance Act (England) lost on an average 28 days of work a year through illness, compared with about 16 days of work 15 years ago."

When you study the longevity figures you find benefits accruing to a great extent to the low-age brackets. After you eliminate infant mortality and maternal mortality and diseases of childhood, and after you adjust the figures for deaths due to pestilence, plagues and lack of cleanliness, your chances of living after you reach the age of 35 are just about the same today as they were 100 years ago. Look at some of the ages of our Revolutionary figures: Benjamin Franklin lived to be 84, Jefferson 76, John Adams 91, John Quincy Adams 81, Washington 67, Madison 85, Monroe 73, Morris 72, Martha Washington 70, John Jay 84, Betsy Ross 84, LaFayette 77, General Gates 78, Gerry 70, Andrew Jackson 78, John Marshall 80, Tom Paine 72.

Sufficient proof, it seems to me, has been gathered to show that this low status of American health is directly related to the condition of the soil in which we raise our foods. Dr. Alexis Carrel, in his *Man, the Unknown* realized the vicious part that chemical fertilizers play in reducing the nutritional value of our foods when he wrote: "Chemical fertilizers, by increasing the abundance of crops without replacing the exhausted elements of the soils, have contributed indirectly to change the nutritive value of our cereal grains and our vegetables."

The problem is recognized too by Dr. Russell M. Wilder of the U. S. Department of Agriculture who in an article in the April, 1943, issue of *The Scientific Monthly,* wrote:

"The soil is poor in certain regions and the products of such soil fail to contain some nutrients that ought to be contained in them. This at present is a major topic of investigation of the Agricultural Research Administration. The chemists will in time be able to correct such abnormalities of the soil. The aim of agriculture until now has been at maximum yields per acre or at size and appearance of the product. The aim in the future should be set at high nutritional values. . . . Experiments performed under my direction confirm the views of many others that the present food supply is most unsatisfactory."

There are interesting facts in this statement. First, Dr. Wilder is a physician. Second, he is employed by the U. S. Department of Agriculture. Third, he says that there are many others who are dissatisfied with our present food supply. Fourth, and a somewhat disappointing fourth, is the fact that he says *chemists* will be the ones to correct these abnormalities in the soil. He might better have said biologists. The chemists have had their chance and failed.

Another bit of evidence showing how the land and our health are directly connected is the "heart chart" issued by the Heart Association of America, which shows that deaths from heart disease are increasing alarmingly in those states that have been farmed the longest. In such areas the land has been "mined." Valuable soil nutrients have been removed in the crops and they have not been replaced by chemicals. Essential mineral elements which ought to be present in our foods are therefore absent.

One of the readers of *Organic Gardening,* G. H. Lane of Eureka, Kansas, writes:

"In my youth, my folks moved to Kansas where the country was just being settled. All food products were grown on virgin soil. People were strong and healthy, except for cases of malaria and a few typhoid. There was but little sickness. People did not know much about sanitation, but lived naturally with no hospitals and no doctors. Large families were the rule. No one ever officiated at births other than one of the neighbor's wives. No canned, pickled, denatured, or chemically preserved foods. No white flour. Only whole grain corn and wheat, ground without processing, plenty of vegetables, wild game of many kinds, and later an abundance of home sun-dried apples for winter use. No insect pests ever bothered the apples, peaches and pears and spraying fruit trees was unknown."

That the vitamin content of our food is dependent on the quality of the soil which mothered it has already been shown by several experiments. In 1926 McCarrison, in experiments with grains at Madras,

India, found that if grown organically they contained more vitamins. This discovery is summarized in the *Journal of Indian Medical Research*.[1]

Mental disorders are terribly on the increase. The National Committee for Mental Hygiene is trying to find the cause, and methods to heal, this serious type of sickness. They have found that mental disturbances "fill more hospital beds in this country than all other disabling diseases combined and that there is a great and evergrowing demand for hospital and clinic accommodations for the care and treatment of sufferers from mental and nervous disorders."

The enormous proportion of men rejected or discharged by the United States Army for neuropsychiatric reasons is an alarming indication of the poor health conditions of modern society. In a recent study made by the New York City Committee on Mental Hygiene, it was reported that 135,500 men from New York City alone were so classified. Only a small proportion of these men can have access to psychiatric treatment—the usually recommended course—because of the expense and time involved, and because of the extremely limited number of psychiatrists.

While it is true that in times of war and stress, the difficulties of adapting oneself to new sets of values and behavior patterns, and to constantly increasing uncertainties and a fundamental sense of insecurity, causes neuroses and mental conflict, is it not possible that malnutrition has something to do with it too?

Another mental health problem—one that may seem far afield from that of food and nourishment—is that defective mentality. Have we, perhaps, too long considered this a matter solely of heredity? Whether food produced from fertile soil can change this factor is a matter that we literally know nothing about. But research scientists are discovering that the "heredity" factor is not all-important as a cause, that many "spastics" (so often classed as "hereditary imbeciles") derive their affliction from a deprivation of oxygen in the brain—in pre-birth stages, or during birth, or from falls or injuries later. A recent report tends to indicate that mental defectives born in intelligent families are the victims of a similar deprivation of oxygen in the brain at some

[1] 14:351, 1926.

time before birth. Is there a possible connection here with a faulty diet in the mother? We do not know—but if research can tell us anything about this, should it not be undertaken?

Dr. Alexis Carrel comments on the connection between food and mental health in *Man the Unknown:*

"Man is literally made from the dust of the earth. For this reason his physiological and mental activities are profoundly influenced by the geological constitution of the country where he lives, by the nature of the animals and plants on which he generally feeds. His structure and his functions depend also on the selections he makes of certain elements among the vegetal and animal foods at his disposal. The chiefs always had a diet quite different from that of their slaves. Those who fought, commanded, and conquered used chiefly meats and fermented drinks, whereas the peaceful, the weak, and the submissive were satisfied with milk, vegetables, fruits and cereals. Our aptitudes and our destiny come, in some measure, from the nature of the chemical substances that construct our tissues. It seems as though human beings, like animals, could be artificially given certain bodily and mental characteristics if subjected from childhood to appropriate diets."

Carrel also remarks:

"It (the organism) is also affected by the deficiencies of the essential physiological and mental functions. The staple foods may not contain the same nutritive substances as in former times. Mass production has modified the composition of wheat, eggs, milk, fruit, and butter, although these articles have retained their familiar appearance. Chemical fertilizers, by increasing the abundance of the crops without replacing all the exhausted elements of the soil, have indirectly contributed to change the nutritive value of cereal grains and of vegetables. Hens have been compelled by artificial diet and mode of living, to enter the ranks of mass producers. Has not the quality of their eggs been modified? The same question may be asked about milk, because cows are now confined to the stable all the year round, and are fed on manufactured provender. Hygienists have not paid sufficient attention to the genesis of diseases. Their studies of conditions of life and diet, and of their effects on the physiological and mental state of modern man, are superficial, incomplete, and of too short duration. . . ."

3

DISEASES OF FARM ANIMALS

Just as man suffers because of the devitalized foods he eats so do farm animals. The situation with respect to animals, however, is far worse. In some outbreaks of farm cattle diseases there is a mortality of 80 and even 90 per cent. Nothing anywhere like this percentage applies to human disease, even though man contracts many diseases from animals, and also eats the meat of diseased cattle and fowl despite federal inspection laws.

Take one disease of sheep. The U. S. Department of Agriculture reports that over 100,000,000 lbs. of lamb are lost every year through a disease caused by nodular worms. The annual value of the meat alone is $10,000,000. $6,000,000 worth of intestines that might have been used as casings for meat products have to be scrapped too.

A similar survey showed that almost half of all the pigs farrowed never reached the market. The debates in the House of Lords, mentioned elsewhere, disclosed that "80 per cent of dairy cattle passing through the markets are diseased in one form or another. The milking life of the average dairy cow is now reduced to two and a half years." It used to be at least double that, and according to Howard, even today, a well managed (organic) herd of Jerseys averages eight lactations in Great Britain.

In the State of Pennsylvania about 13 per cent of the dairy herds are infected with the dreaded Bang's disease. In one month recently Government subsidy payments to Pennsylvania farmers for the slaughter of diseased cattle amounted to $53,000. It is estimated that the total annual loss from this disease is $50,000,000. Bang's disease causes sterility in cows, and undulant fever in man and is one of the greatest difficulties dairy farmers today have to contend with. The Virginia Agricultural Experiment Station recently found, in a special survey, that 25 per cent of the 844 beef cows studied could not produce calves.

In the meantime the farmer has to spend about $27 upkeep on each cow per year whether it calves or not.

Mastitis in cows is cutting production by 20 to 25 per cent in many parts of the country. This means the difference between profit and loss on many farms.

The situation of general farm animal disease has reached such a point that the Veterinarians (American Veterinary Medical Association) at their August, 1944, Convention at Chicago proposed the establishment of animal disease and mortality statistics registration similar to that of the general population.

The approach to the problem by animal doctors and the scientists is, of course, curative rather than preventive. They use medicines, vaccines, serums, the new sulfa drugs and penicillin, if they can get it. In 1939 sales of animal and poultry disease remedies amounted to $81,000,000. But this is a negative approach. These diseases are merely an indication that something is causing trouble in the animal. Curing a cow with a dollar's worth of some medicine purchased at Sears Roebuck or the drug store is not going to remove the cause. These diseases are warnings by Nature that something is wrong and that it is necessary to start rebuilding from the soil up—to feed animals quality food raised in quality soil.

Dr. William A. Albrecht, head of the Department of Soils of the University of Missouri, in an article in *Organic Gardening* warned:

"Evidences of declining soil fertility are seen in the greater number of deficiency diseases among our farm animals. Veterinarians are constantly faced with increasing cases of strange animal ailments for which no specific body weakness or visible physiological cause can be found. Eye ailments, a tendency to blindness, bad gaits, rounded back lines, inferior condition, poor feeding progress, and even debility and death can be traced to deficiencies in animals' nutrition.

"Confined as they are, animals are limited in their feeding by the soil fertility of the farm. Supplements such as proteins contain too little of the minerals needed. And when animals protest by rooting up the very earth or climbing the fence in search of better forage, such actions are met by rings in the nose or cumbersome yokes about the neck. Human serfs have been prohibited in this country for about 75 years, but animal slaves can be seen on any trip throughout the country. Too often they are almost in a state of starvation because their master, the farm owner, neglects his soil fertility."

Some of this near-starvation can come from unwittingly feeding poison-loaded foods. In *The Spectator* (London), October 17, 1941, there appeared this item by T. H. Sanderson-Wells, Chairman of the Food Education Society:

"A dig-for-victory plot on the edge of a golf-links was limed, planted and treated with artificials. Luxuriant heavy-green cabbages, sprouts and other vegetables resulted. To increase the family meat ration, part of this crop was fed to rabbits, who ate without relish, became apathetic and smelt unpleasantly. When later, grass mowings were substituted the rabbits ate voraciously and became vigorous and sweet-smelling.

"A correspondent writes: 'Cabbages and sprouts grown too fast with nitrate and phosphate are a curious wrong colour. If over 50 per cent of the greenstuff given to rabbits is of this sort the rabbits die. Permanent pasture dressed with phosphate produces a luxuriant field. If the phosphate goes beyond a certain point the field takes on an unnatural green, and is deserted by wild rabbits.'

"Salesmen use this fact as a recommendation. One told me: 'Use any soluble phosphate fertilizer and keep the rabbits away.' Another said: 'Use enough nitro-chalk and you will get big greens that rabbits will scarcely touch; if they do, they die.' Animal instincts may be sound guides to food values, which are actually soil values, because food is nothing more than the 'conveyancing agent' or 'agent of transfer' of the soil's qualities into the bodies of man and beast; land in good heart supporting bodily health, vigour and stamina; poor unbalanced soils producing ill-health and debility. For instance, the liking of birds for hips, haws and many other hedgerow fruits has led to the discovery that these contain high concentrations of mineral salts, vitamins, and other essential food elements."

In *The Gardener's Chronicle* (England), appeared the following item on April 11, 1942:

"At Marsden Park in Surrey in 1939, the following results from feeding trials were obtained. 'Poultry, pigs, horses and dairy cows were fed separately on a grain ration raised from fertile soil and compared with a similar one purchased in the open market. In all cases the results were similar. The animals not only throve better on the grain from fertile soil, but they needed less—a saving of about 15 per cent was obtained.'

"Resistance to disease markedly increased. The infantile mortality in poultry fell from 40 per cent to 4 per cent. In pigs, troubles like scour disappeared. Mares and cows showed none of the troubles which often occur at birth."

This happened on the farm of Sir Bernard Greenwell where the soil had become extremely fertile following the exclusive use of humus as the fertilizing element.

The Earl of Portsmouth mentions his experience with farm animals, in his speech in the House of Lords, previously referred to:

"When I started to farm some twenty years ago I was thoroughly up to date with modern ideas, but gradually by trial and error—far more often, I may say, by error than by success—I revised all my previous notions. I found again and again that, despite what analysis proved, the quality of bought food was very low compared with the very genuine food value in my own home-grown foodstuffs. For instance, in a comparison between protein in beans and in oil-cakes, my beans won every time. The same thing was found in home-grown oats. The old analysis showed their food value to be very low, yet practical experience in feeding home-grown oats to my own cattle showed the value of the oats to be much higher than anything that could be bought, except the most expensive foods. I found that my animals had a bloom. That experience brought me back to the necessity for consulting nature instead of trying to beat her. I have come, therefore, through the very hard force of circumstance and by practical trial on my own land, to believe that there is more in the way we treat the soil than there is in any methods of trying to get the maximum out of the soil by artificial means."

Lady Balfour's experience with young pigs is very similar:

"These animals if kept shut up in sties or houses, which is the only way that they can be kept by allotment holders, are very subject to white scour which attacks them when they are about a month old. I have proved to my satisfaction that if these young pigs are kept supplied with fresh soil from fertile land, that is, rich in humus and where no chemicals have been used, they do not suffer from this trouble. The soil should first be given when the pigs are about a week old, and should be continued up to the sixth week. You would be surprised at the large quantities that these baby pigs consume. Now here is the interesting thing. If the soil be taken from land that has received the usual dressings of chemical manure, and no compost, it is quite ineffective either as a preventive or cure for this complaint."

Animals have a way of showing that all is not well with their food. Pigs have been known to chew their wooden troughs all to pieces where there is a deficiency in their foods. Sows often eat their young because some elements in their food supply are wanting. Horses chew out parts of their stall doors or mangers under similar conditions. Bulls lose some of their breeding powers when their food does not contain sufficient vitamins and minerals. Chickens eat their own eggs. And so it goes. A strongly endowed food means healthy animals.

The cause of many cattle diseases lies in the devitalized food they get, and, the worst kind of food is the kind that is specially grown for farm animal feed purposes. Tremendous acreage is devoted to the production of crops that find their way into the concentrated, factory-processed feeds purchased by many dairy farmers. More chemical fertilizers, no doubt, are used on such crops, and less—if any—stable manures. By costly advertising campaigns these large feed concerns drum it into the heads of farmers that they cannot produce good milk without these concentrates: that they will give their cattle health.

Some of these concentrates, and other cow and chicken feeds, contain by-products of the soap and oil industries and sometimes condemned food substances. It doesn't take much grey matter to see that the substitution, for these purchased foods, of crops raised on the farmer's own land, made rich by yearly applications of humus, would considerably raise the health level of his cattle. Many men dairy as if they were running a milk factory. They have many cows and practically no land. They depend on purchased foods. They even sell the cow manure as a by-product. Theirs is the poorest kind of milk and the poorest kind of manure.

Government experiment stations have been working with various fillers for cattle feeds. Synthetically produced urea is gradually coming into use with the blessing of many of these stations, yet Massachusetts State College found that when it was used at a rate of 60 pounds to the ton for two lactations, there was a reduction in the milk and in the cow's condition during the second period. It was found that the cow was the only farm animal able to stand urea because it has four stomachs or compartments. One of these, the *rumen,* contains large numbers of bacteria which break down the urea. It cannot be fed to horses, poultry, hogs or young calves. When urea is in the ration care must be taken to exclude soy and other beans, because they liberate nitrogen from the urea, which may cause disorders in the cow.

Now they are working on sawdust to be processed for use by yeast cells to be fed to cattle. Steel tanks are filled with sawdust and dilute sulphuric acid is fed down through the mass. The liquid which comes out at the bottom is worked on by yeast cultures. (Remember what has been said previously about sulphuric acid. This chemical seems to pop up everywhere in agriculture although it is a violent poison.) We are not finished, however, because in the process, mineral salts have to be added, usually ammonium sulphate, our old friend, which is used to kill earthworms. The end product is then fed to cattle.

With the aid of concentrated feeds and by selective breeding, cows are being developed that are giving enormous yields of milk, but as someone has said these are not cows and their milk is not milk.

The cow has a tough job. She must process much larger amounts of feed into milk than nature intended and she must still produce a calf every year. Is it any wonder that she is delicate and takes sick easily. It is a simple matter for her reproductive system to get out of kilter. Beef cows do not seem to suffer from anywhere near the amount of sickness that milk cows do.

Pathfinder Magazine [1] describes a new method of getting the cow to give more milk by feeding her synthetic hormones. "If there's a catch," says *Pathfinder,* "to the new method of getting more milk from cows, it may lie behind tests showing the synthetic hormone decreases vitamin C content of milk by one-third. An equal decrease in the vitamin C content of a cow's blood might impair her breeding value, since vitamin C is closely connected with fertility in cattle."

In the *History of Randleigh Farm,* Drs. Ernest Scott and L. A. Erf of Ohio State University tell of experiments where some rats died when fed on milk from high-producing cows, and others recovered

[1] June 26, 1944.

when fed on milk from an ordinary herd, and that "experience has taught breeders that the calf of a high-producing cow does not thrive on its mother's milk but remains weak and lifeless. Such calves when placed upon milk from cows of lower production soon develop normally." Amazing, but true; and what does such milk do to us?

Dr. R. H. Smythe writing in the *Veterinary Record* (1942), raised the storm-warning signals when he said:

"Thirty years ago, the percentage of infertility in bovines was negligible; today it threatens to dislocate or even destroy the dairying industry. It is significant that the breeding index of heifers has fallen enormously of late years, even on farms where nutritional problems do not exist, and where the incidence of concurrent disease is low. The purpose of these notes is to stress the opinion held by the writer that with the possible exception of brucella infection, herd infertility as well as the failure of individual animals to breed is in great measure based primarily upon physiological rather than upon pathological causes, and that selective breeding, with the object of constantly stepping-up milk production, is inducing the development of lethal factors.

"It is a mere question of time before the dairy cow and her fellow victim, the laying hen, become automatically extinct. By a routine adoption of our present methods of treatment we are merely temporizing in maintaining a degree of fertility which may suffice for the moment. Incidentally, we are helping to fix a strain of cattle which will depend more and more upon artificial breeding aids in each succeeding generation."

May there not be some connection between this infertility and the inability of plant strains to reproduce themselves, mentioned earlier? Talking about artificial breeding aids, in the last few years the artificial insemination of cows has been gaining wider and wider usage. It might be safer if this practice could be tried in the experimental stages for a period of 30 to 40 years. But thanks to the "trial-and-error" methods of modern science, artificial insemination stations are being established at every cross-roads junction. If there *is* going to be a repercussion later, it will be too late.

Nature has planned her own method for the perpetuation of the species. With artificial insemination man is initiating a practice that may be fraught with deep biologic significance far beyond our present capacity to understand.

4

PLANT DISEASE AND INSECT PESTS

Both of these seem to be increasing just as fast as are diseases of farm animals.

In order to cope with this problem—it would help a great deal if the farmer and gardener tried to study cause and effect. First of all, what causes plant diseases? Most of them are caused by bacteria, fungi and other microscopic disease-producing organisms. Do these organisms directly bring about the disease? The answer probably is *yes*, but then the question arises, *why do these organisms attack certain plants and leave others alone?*

Let us consider a human being. The mouth of an average person, even after he has cleaned his teeth, teems with bacteria. When bodily resistance gets low some of these germs or viruses may bring on a cold. But some people can harbor quantities of identical disease-causing organisms without ever contracting a cold. It is said that practically everybody carries tuberculosis germs in the body—but comparatively few succumb to the disease. The plant world presents a parallel situation.

In the human body, resistance is chiefly built up by eating a varied diet of good food. Even though our food today is more or less unsatisfactory, a person receiving a rounded diet is healthier and better able to resist disease than one who lives on an unbalanced diet (other things being equal). The food of the plant comes from the soil where it is predigested before ingestion takes place through the roots.

We have shown that earthworms, bacteria, fungi and other microorganisms play a prominent part in breaking down organic matter and preparing it for the roots; it has been demonstrated beyond question that the use of strong chemicals inhibits the action of these microbes:

and that where practically no organic matter is used, and dependence is entirely on chemical fertilizers, the plant is apt not to be healthy. Professor Waksman in *Humus* is in agreement when he writes:

"Plant deficiency diseases are usually less severe in soils well supplied with organic matter, not only because of the increased vigor of the plants but also because of antagonistic effects of the various soil micro-organisms which become more active in the presence of an abundance of organic matter."

In the *Rhodesia Herald* of September 4, 1942, there is published a confirmation of the value of organic matter in giving plants health. The following is an extract:

"Some years ago Mr. S. D. Timson, Assistant Agriculturist, noticed a garden in which the vegetables were strong and the flowers bright and vigorous. He was surprised to learn that three years earlier cultivation had been almost abandoned because of the heavy infestation of eelworm. The excellent conditions he saw followed a good dressing of compost. He immediately began to observe the results of compost in regard to eelworm, make practical tests and induce farmers to experiment. Once the inquiry was begun evidence began to pour in.

"At Darwendale, Mr. O. C. Rawson had applied five tons of compost per acre to infested tobacco land. In the first year there was a reduction of eelworm, and in the second year, without a further application, the eelworm disappeared. Other tobacco farmers began to report similar experiences. . . . The compost, of course, was applied for its fertilizing value, and the consequences on the eelworm population were a surprise.

"It now seems that the same effect has been noticed in other countries where compost was being tested out. In Ceylon, members of the Department of Agriculture were able to report last year that the most promising method of ridding the soil of eelworms was one they had only recently discovered. It had been found that if large quantities of organic material, such as compost, green manure, or cattle manure, were added to the soil, the population of eelworms was greatly reduced. Examining into the causes of this effect, their conclusion was that it was due to the increase in the soil, following the addition of organic matter, of organisms like fungi, and insects and other nematodes which preyed on the eelworms. . . . From Ceylon it is also reported that as research work on plant diseases progressed it was becoming increasingly evident that many of the soil-borne diseases of that island would not be so severe if more care were taken to increase the amount of humus and organic matter in the soil.

"Plants grown in well-cultivated soils, well supplied with organic matter, are not affected by many plant diseases to the same extent as those which are grown in the poorly-cultivated unmanured soils."

We have cited previously Dr. J. Sidney Cates's experience showing that root-rot disease in cotton was eliminated by the simple expedient of applying stable manure. Much of our cotton is grown with a 100% chemical fertilizer application in a one-sided monoculture. The curbing of the root-knot parasite on Hawaiian pineapple plantations was accomplished in the same manner—substantial use of organic matter.

Sir Albert Howard wrote in *Organic Gardening* in 1942:

"In 1934 I purchased a house, the garden of which was completely worn out through no fault of the previous owner. It was a veritable pathological museum—the fruit trees, in particular, were smothered with every kind of blight. Steps were taken to convert all the vegetable wastes into humus with the help of stable litter. Even after one year the pests began to retreat. In three years all had disappeared, the wooly aphis on one apple tree being the last to leave. During this period no insecticides or fungicides were used and no diseased material was ever destroyed. It was all converted into humus.

"One further interesting result must be mentioned. I grew in heavily composted soil a collection of Royal Sovereign strawberries, badly infected with a common virus disease, alongside some healthy stock. This year, 1942, the strawberries raised from these two sets of plants were about the best I have ever tasted. I found no trace of the virus disease. Similar results have been obtained by several of my correspondents."

Eve Balfour in the *Compost News Letter,* writing about her experience with six acres of vegetables, shows once more that compost-grown crops can resist pests remarkably well:

"You know what a difficult season it has been for that class of crop, what with the drought, the fly and the caterpillar.

"My seedlings were grown on fertile soil, and were planted out on the ridge, compost being first put under the ridges. The weather was dry when they were put out, and in a short time all the plants were badly attacked by, first fly, and then caterpillar. I thought I had lost the whole crop; but I did nothing about it. In a few weeks, when the roots reached the compost, the plants began to recover, and, despite no rain, they not only grew well, but began to master the pests. About six weeks ago you could see side by side the completely recovered and the partially recovered and a few still badly attacked. Now, you would hardly believe that there had been a caterpillar or a fly in the field. My buyer says they are the cleanest greens he has seen anywhere this year. The flavor is superb, and the color has been good throughout. He himself has lost all his cauliflowers and most of his cabbages; and has spent a fortune on sprays and insecticides. Result: he is the latest convert to compost. I have often grown these crops on farmyard manure without artificials; but, even if otherwise healthy, they have always before been bothered with fly in a dry season.

"What interests me so particularly about this experience is that the crop was attacked and recovered, without any treatment. I find this more impressive than if they had not been attacked."

One more illustration seems worth reporting. F. C. King in his book, *Gardening with Compost* relates the following experience with raspberry canes:

"About twelve years ago I planted some raspberry canes of the *Lloyd George* variety, which were stunted and so badly infected with virus disease that the owner was only too pleased to give them away. Below I give the yield obtained from 100 yards of these canes during the last ten years:

1934—115 lb.	1939—137 lb.
1935—128 lb.	1940—107 lb.
1936—273 lb.	1941—118 lb.
1937—238 lb.	1942—249 lb.
1938—153 lb.	1943—214 lb.

"During the autumn of 1936 I commenced alterations in the pleasure grounds of Levens Hall, which kept the small staff fully employed and in consequence the raspberries were neither cultivated nor mulched until the autumn of 1940 when I cut out all the old fruiting canes, tied in the new canes, and cleaned the plot of weeds. The rows were mulched with compost and the space between the rows forked up. Early in April another mulching was given, and the space between the rows forked again. This method has been repeated each year and has resulted in a heavier yield of fruit—much heavier in fact, than my figures show, as a good proportion of the crop each season could not be gathered on account of bad weather.

"I believe my modest test has clearly proved that a return to the use of compost must be made if we are to retain good cropping qualities of our stocks of raspberries over a reasonable period of time. I have been told, however, that I have not succeeded in curing the virus disease but only in masking the symptoms. My reply to this objection is that I am satisfied that by the use of compost I have secured good crops for the past 10 years from canes, which, when planted, were incapable of producing fruit, and even if compost does no more than mask the disease, the fruit my canes have produced would do credit to any gardener and the practical results I have obtained seem to make the difference between masking the symptoms and curing the disease relatively unimportant."

When insects come to ravage plants it is easy for the gardener to get out his spray-gun and saturate his garden with lethal doses—lethal for the bugs and also for the helpful life in the soil. You can't blame him because he has been systematically propagandized and has come to believe that it is the only thing to do.

Nature, however, always can be depended upon to have a reason for her way of doing things. We don't put poison spray on bees. We know they are beneficial. Do all other insects serve some particular function in Nature's general scheme? Insect pests are not the cause of plant disease. They are an agency of Nature which point out that something is wrong with a plant—either it is imperfectly grown or the soil on which it feeds is not properly fertile. Nature has provided predators to remove the unfit. It is part of her scheme for the perpetuation of the species that the best survive and the weaklings are removed. There are, of course, exceptions to this rule deriving from man's interference with Nature's scheme. In the case of introduced species of pests such as Dutch elm disease, chestnut blight and many other diseases of insects, plants and even animals, factors which normally keep them in check are missing from their new environment and time must elapse before a new balance is achieved. Artificial controls can hardly be avoided. Again, it is not "natural" for pure stands of certain species to occupy large acreage to the exclusion of other species. Such stands often provide an artificial environment for certain pests to become established beyond their ordinary numbers. Trouble of this sort can best be avoided through rotation of crops.

The operations of Nature work out in as nearly perfect a manner as is practical in the involved and complicated world of living things. The processes of decay set in as soon as organic matter dies. When a leaf is on a tree it is living and growing. As soon as it falls off, it begins to decay with the aid of a host of Nature's helpers. The same thing applies to any living matter. Can you conceive of a world where matter could not decay?

Now let us return to the leaf on the tree. Sometimes a blight hits it. It seems to start decaying before it is dead. Or it may be eaten by some insect that does not seem to eat leaves from other trees of the same species. When you seek the cause you will find that the tree is sick. Isn't sickness the beginning of death, the beginning of the processes of decay? Have you ever dumped table garbage on the ground or in a hole? In a day or so maggots begin to work on it. This is an important rule of Nature. These maggots will not feed on live matter. By and large, in the same way certain insects prefer diseased tissue or weak individuals. Nature has apparently given them a perverted taste for their scavenger function, for they do not seem to relish the taste of a healthy plant.

Take the case of the apple tree borer. It is the weakened or sickly trees that are seized upon by these pests. Healthy vigorous individuals seem to throw them off—probably because the eggs which are laid in the bark are quickly surrounded by new bark. Much investigation remains to be done in this whole field. Is it possible that healthy plants have a taste or smell that repels these predators?

We had an experience with Japanese beetles and grapevines that seems to bear this out. On our farm when we purchased it, there was a very old grapevine which was stripped by these beetles every year. Last year we planted a young new vine near the old one. We put in a good application of compost and a heavy mulch under each vine. The leaves of each intertwined, yet the beetles still worked only on the leaves of the old vine. They left the new plant alone, even though its leaves were right in among the others. Here or there could be seen a little bite, but evidently one taste was enough. The chances are that the old vine was too far gone to rehabilitate, even with compost. Or it may take two or three years to revive it. We shall see. In the meantime, the Japanese beetle seems to distinguish in its taste two kinds of leaves of the same species. The new vine was young and it responded to humus more quickly than the old one.

Many old farmers will tell you that they had much less trouble from insects and plant diseases when they were young. I have asked dozens of them and always got the same answer. It seems that insect infestation and plant diseases are concomitants of poor cultural methods, and are aggravated by the use of chemical fertilizers. The potato blight as a regular phenomenon developed into a major problem just about the time that strong chemical fertilizers came into general use. Before that, potato blight seems to have been the result of abnormally rainy seasons—severe and prolonged enough to compact the soil—keep it waterlogged and thus prevent proper aeration.

You never have "perfection" in Nature. There will always be insects, some useful to man, some rivals of man. We grow potatoes in an organically treated soil yet we do have *some* potato bugs, but not many. Why get excited and tear our hair over a handful of potato bugs. They do very little harm, much less by far than the polluting of the plant and soil with poison sprays. A thing really to get excited about is the new potato disease that recently attracted national attention. Roger W. Smith in an article in *Organic Gardening,* sizes it up for us:

> "It consists simply in the formation of a hollow tuber. Unfortunately, this hollow tuber makes possible a new focus of infection for other more virulent diseases. Much investigation followed its discovery, and although it has not been openly admitted there seems little doubt that the growth of these faulty potatoes was caused by 'improper moisture conditions.' Actually, this really means that the potatoes were grown in poor soil and their growth encouraged by the use of large quantities of artificial fertilizers and floods of water. You realize instantly that potatoes 'blown-up' to such an extent that their centers were hollow could have little actual food value."

In connection with plant disease one is often cautioned to burn up diseased plant residues to prevent their infecting new crops but

Howard advises that if properly composted it may safely be used again on the land. He cites as interesting example:

"A few years ago Mr. A. R. Wills, of the Tadburn Nursery, at Romney, Hampshire, found that the use of chemicals in his tomato houses led to wilt disease. Three houses—three acres in extent—became badly infected. On my advice he composted the diseased haulms (stalks) of these three acres, applied the compost to the same houses and grew a new tomato crop. This was free from disease. He then took up the Indore Process and gave up the use of artificials. His nursery has never looked back. It now enjoys a high reputation for quality and there is a great demand for the produce.

"This pioneering work was followed by many other examples in which composted vegetable wastes, yielded by plants attacked by insect, fungus and even virus diseases, gave rise in all cases to perfectly healthy crops."

Many other examples could be given but the most graphic illustration of this principle involves a case where Sir Albert Howard was called in to witness a peculiar phenomenon. In certain cultivated plants certain rows only were attacked by aphids. On checking he found that the soil under these rows was hard-packed. For some reason, wherever there were no aphids, the soil was loose and porous. Here is one reason why compost permits plants to grow relatively free of disease. It creates a loose, porous soil, well aerated.

To keep a check on the natural activities of insects, bird-life should be encouraged. Many farmers today are removing hedgerows and stone-wall fences, where small brush grows, to increase the size of their fields and to save time in cultivating. By destroying these wild-life and bird havens, the balance of Nature is upset. The birds go elsewhere, removing a powerful curb against the abnormal increase of destructive insect life.

The *Eastern States Cooperator* brought this matter up in a recent issue as follows:

"With the hunting and trapping season approaching it is a good time to consider an important, often neglected, asset of a well-run farm. The wild-life population is a good indication of whether or not sound farming practices have been followed since we have now learned that soil conservation, erosion control, sustained water supply, soil fertility, good timber, are all bound up with the wildlife that lives on the land and that prospers if the land prospers and disappears if the land is wastefully managed.

"How often we hear old-timers talk of the good old days when grouse and squirrel and rabbit and coon and muskrat abounded on every farm, when there were no English sparrows or starlings and the barns, orchard and shade trees were full of native songbirds. They often say they didn't know about insecticides then and didn't need them. They had no strict hunting laws and thought they didn't need them.

"But with more and more land clearing, wood cutting, brush burning, over-grazing, uncontrolled hunting and trapping, wildlife has came to such a low point on the average farm that it is only a memory of the past—along with the horse and buggy and other good things of an era never to return again."

The natural border hedges of brush and trees between fields are not only valuable as bird havens, and thus for insect control; they are very effective as windbreaks in keeping the drying winds down and the temperature of the ground up, a most important point for many crops; important too for the land in preventing wind erosion. Large corporations have entered the field of agriculture and cultivate thousands of acres. The first thing they do is to cut down the hedgerows so that their tractors can operate more efficiently. This is a very bad practice.

I do not need to emphasize the great role birds play in insect control. But in this connection, I want to quote from the redoubtable Horace Greeley, not only for what he says about birds as insect-eaters, but for his altogether valuable advice on good cultural techniques as a means of controlling insect pests—chiefly that of crop rotation. His advice, alas, was not taken at the time he gave it, for beginning about 1870, when he wrote, the plantation, one-crop system of farming, get-rich-quick style, became fashionable in the North and Middle West, extending the soil-robbing culture until then largely restricted to the cotton and tobacco plantation areas of the South. Wrote Mr. Greeley:

"I would be far below the mark if I were to estimate the average loss of the farmers of this country from insects at one hundred million dollars per annum. In my neighborhood the peach once flourished, but does so no more, and cherries have been all but annihilated. Apples were till lately our best crop, but worms take half of them, and sadly damage what they do not destroy. Plums we have ceased to grow; pears are stung or blighted; even the currant has its insect enemy.

"We must fight these paltry adversaries more effectively or allow them to drive us wholly from the field. In this I have no doubt that the best allies in this inglorious warfare are the birds. They will save us if we do not destroy them. Whenever clouds of birds shall habitually darken our fields in May and less notably through the summer months, we may reasonably hope to grow fair crops of our favorite fruits.

"I hold the multiplication of insects and their devastations are largely incited by the degeneracy of our plants caused by the badness of our culture. I presume that wheat and other crops could not be devastated by insects if there were no slovenly, niggard, exhausting tillage methods used. But when the fields of western New York were first tilled there were few insects; but after crops of wheat had been taken from those fields until they had been well-nigh exhausted of crop-forming elements, we began to hear of the desolation wrought by insects. I believe that we should have heard little of insects there had wheat been grown on those farms but one year in three since their redemption from the forests.

"But whatever might have been, the Philistines are upon us, and we are doomed for at least a generation to wage a relentless war against insects, multiplied beyond reason by the neglect of our predecessors." (That was in 1870)

It seems in recent years that our valuable allies the toads are disappearing from farms. Is it possible that the chemicalized soil is distasteful to them? After a rain they used to appear by the hundreds. The toad is extremely useful to the farmer in keeping insect pests in check.

I mentioned this fact to a friend at Quakertown, Pa. Her eyes opened wide. "Why, that makes me think," she said. "Near our house we used to hear the croaking of what seemed to be thousands of frogs. Then one year we rented two acres to a man who planted them in tomatoes, and who used plenty of chemical fertilizer. We never heard any frogs there after that."

The planting of cover crops encourages multiplication of quail who nest in the tall grass. Quail are a factor in keeping nature in balance and an asset to any farm.

The insect menace is beginning to assume colossal proportions. It is estimated that the annual damage amounts to two billion dollars per year, despite millions of dollars spent on research. Over 4000 Government workers are continually at work on this problem. The scientists admit defeat, in that they say that large quantities of destructive insects will always be here to plague us, and that our only means of combatting them is with poison sprays. Possibly a change in our system of research is indicated.

5

POISON SPRAYS

Orthex, Pulvex, Ant-X, Apex
Curex, Go-Nex, Larvex, Japex
Bordex, Bindex, Fulex, Fly-X,
Foamex, Protex, Pathex, Pyex.

Baxco, Bim, Bif, Bindarine,
Calso, CopRote, Calogreen,
Flea-not, Fly-Foil, F. & I.,
—That's the stuff will make 'em die.

The above verse by Anthony Standen appears in his book *Insect Invaders* in which he states that he has come across 953 different trade names of plant insecticides or sprays. This is a well-entrenched industry that will fight hard to continue earning profits.

Poison sprays, although the "easiest way" of killing insect life, are extremely dangerous. On the night of August 7, 1944, about 1200 sheep were killed by the fumes of a poison spray from an olive grove in the Saucelito district of California. The wind caused the fumes to drift. In the same manner many inexperienced gardeners handle sprays carelessly with attendant dangers to plants, to animals and to themselves.

The farmer's health can be affected by close contact with poison sprays, especially if he inhales them. The *Journal of the American Medical Association* for December 5, 1943, relates the experience of two doctors, Herbert H. Kilgore and Paul S. Rhoads, with such a case. The farmer in question had been handling lead arsenate sprays in orchards and vineyards for thirty years:

"In retrospect," the two physicians say, "the patient recalled that as long as five years before he had attacks of nausea and a quivering sensation of his muscles which kept him awake and restless at night but did not bother him while he was working. These symptoms were most noticeable the night following the use of the spray and decreased in intensity each night until the spray was again used, at which time there was an exacerbation of the symptoms. For two years he had had conjunctivitis (inflammation of the membrane lining of the eyelid and the front portion of the eyeball), photophobia (abnormal intolerance of light) and blepharitis (inflammation of the eyelids) each time he had used the spray. . . ."

Further regarding this case, a special bulletin of the American Medical Association says:

"Laboratory analysis of his urine, hair and nails revealed arsenic trioxide. Treatment included intensive vitamin therapy, especially with reference to the vitamin B group, as well as physical therapy in the form of heat and massage. Seven weeks after treatment had been started, a chemical analysis revealed a considerable reduction of the arsenic trioxide content of his hair and nails and he had gained 25 pounds."

Nothing is said here about attacking the *cause* of the disease.

Many lawns that have been treated with arsenate preparations to kill off various kinds of grubs contain dangerous residues of such poisons. This should be taken into consideration when such land is used for vegetable gardens. Professor Firman E. Bear of Rutgers University notes in a letter bearing on this: "There is abundant evidence that plants growing on soils that have been treated with arsenate contain more arsenic than do the same species when grown on untreated soils." The Boyce Thompson Institute of Plant Research (Yonkers, N. Y.) advises that vegetable plants should not be grown on soils that have been treated with arsenic for five years after such treatment.

Leonardo da Vinci in an experiment long ago showed that arsenic placed at the base of fruit trees actually found its way into the fruit and in an amount that was considered dangerous. Pirotta experimented in Italy along this line in 1906 and found that a dilute arsenical spray on plants resulted in an increase in the amount of arsenic contained in the plant.

It requires an expert knowledge of sprays to handle them without injuring growing things. *Horticulture* of June 15, 1944, carried the following item:

"Dr. Paul E. Tilford has reported to *Arborist's News* that a light application of dusting sulfur to the foliage of Viburnum carlesi caused black spots to develop on the leaves. The leaves dropped within a few days after the spotting appeared. Plants of this variety which may be growing near a rose garden are often so severely injured that they may be defoliated by drifting sulfur dust intended for the control of black spot on roses."

The *Country Gentleman* of March, 1944, carries a similar item:

"Injury from arsenic spray used in controlling circulio has been encountered by peach growers, particularly in the eastern half of the country. The damage shows up in cankers on the growth made during the current year as well as on the one-year-old wood, and the leaves also display marginal burning. In many cases the cankers are large enough to girdle the shoots or small branches so that by harvest time trees may have much deadwood."

One can cite examples by the dozen in print of admonitions, cautions, and cases of disastrous happenings if directions are not observed to the dot.

A vexing difficulty encountered in the use of poison sprays is the perverse nature of the insects themselves. Drat them, say the chemists, they will not act according to formula. The little devils develop a resistance to the poisons.

This is shown in Bulletin No. 4 (1942) of the California Department of Agriculture, which says:

"Tartar emetic was first used in combination with sugar for control of citrus thrips about three years ago, and for the first two seasons it gave spectacular control. Early this year a research entomologist submitted for our examination a sample of tartar emetic that had failed to afford satisfactory control of citrus thrips in San Fernando Valley. Analysis did not differ significantly from those previously made of a number of similar products. Further investigation revealed that even repeated applications of tartar emetic sprays do not now afford control of thrips on citrus in this locality. Information also indicates that thrips exhibited a similar 'resistance' in the Porterville area in Central California, but were satisfactorily controlled in other areas."

Lord Portsmouth in the debate in the House of Lords on February 2, 1944, illustrated the same point:

"Compare that with what you could have seen before the war on a visit to one of our research stations in England. There you could have gone into the laboratory and have had explained to you a new form of spray for poisoning aphids on apples. The history of the need for that new spray was that round about 1900, when apples were being sprayed with arsenic to keep them free from various attacks, several people died of arsenic poisoning through eating apples, so that a law was introduced prescribing the maximum amount of arsenic that could be sprayed on apples. But about 1935, 200 times the legal amount of arsenic was being sprayed on apples which had to be washed afterwards with some napthalene solution. The various parasites which the arsenic was supposed to kill had developed a resistance to arsenic, and although the strength was increased 200 times the parasites were still there. Therefore it became necessary to develop a new form of poison. But on those farms in Holland no sprays of any sort, apart from certain vegetable sprays, were ever used."

The farms in Holland that Portsmouth is referring to are run on the organic plan, using no strong chemical fertilizers. A bulletin of

Arthur D. Little Associates, engineers of Cambridge, Mass., issued during 1944 said, along the same lines:

"Black scale and California red scale have become resistant to hydrocyanic acid fumigation of citrus trees. Larvae of the codling moth learned to resist arsenical sprays, and primary screw-worm larvae, an animal infestation, to survive the once deadly phenothiazine. The period of acclimatization varies from two to twenty-five years, and a remedy is to alternate insecticides over a considerable period; citrus trees, for instance, are treated with hydrocyanic acid one year and an oil spray the next, thus confusing the parasites. Fortunately no insect is expected to reach the ultimate in adaptability by learning to resist any regular sequence of attacks."

Who knows, give these insects a little longer and they may learn to become acclimated even to alternations.

Another dangerous factor in poison sprays is that the poison kills off the natural enemies of the destructive insects. Hundreds of cases in point could be cited similar to this, a paragraph from an article "Advances in Entomology," by E. R. Woodbury, in *Industrial and Engineering Chemistry*[1]:

"Schoene, Virginia Agricultural Experiment Station, Blacksburg, finds that the white apple leafhopper is usually much more abundant in sprayed than in unsprayed orchards, the explanation being that applications of lime sulfur and Bordeaux mixture tend to hold in check a fungus parasite on the leafhopper. Other instances of the destruction of the natural enemies of predatory insects are given."

The *Westmoreland Gazette*, England,[2] has something interesting to say on this same point:

"The most curious phenomenon has been the plague of greenfly, which has played havoc with many fruit crops, both hard and soft. 'Curiouser and curiouser' is the fact that the aphid pest appears to have been most prevalent and destructive in gardens where regular and systematic spraying has been indulged in. Various washes in many orchards seem to have provided sustenance for the pests, while in other cases, where the owners have used neither sprayer nor pruning knife for years, fruit crops have been normal. These circumstances,

[1] January 10, 1939.
[2] August 12, 1944.

though exasperating to those gardeners who have endeavoured to be 'up to date' and have used all the recognized scientific measures to combat pest invasions, give rise to an intriguing question: Are we interfering too much with Nature?"

During the early part of 1944 a new "wonder product" was announced, called DDT, a military development of World War II, hailed as a wonderful delousing agent for soldiers. For a long time it was kept secret, until the facts about it leaked out after it brought into control a virulent epidemic of typhus in Italy. Originally it was used by a Swiss firm to control crop pests and is now being widely advertised as a safe insecticidal spray, harmless to humans. But the U. S. Government and the American Medical Association seem to think otherwise. Doctors of the U. S. Public Health Service find that it acts like carbolic acid and in experiments it has killed animals. They state that it is a definite health hazard.

With regard to this product *Science News Letter* of Aug. 5, 1944, says, "They may at the same time destroy both useful and harmful agricultural insects. They may rid your dog of fleas, but insidiously, and perhaps fatally, damage his liver or paralyze him through nerve damage. They will rid your home of mosquitoes, flies and vermin, but the price may turn out to be high in human health and life."

Other agricultural insecticides are definitely being advertised as deadly to insects but harmless to man. This is rather strange because some of these products contain poisonous cyanide. Others are made up of organic poisons, which are praised as being safe, since it is claimed they leave no poisonous residue in soils. They do, of course, kill beneficial insects as well as harmful ones, and further research is needed before they can be definitely pronounced safe.

Besides poison sprays there are other practices in connection with insect eradication that might be questioned. Don't be too quick to adopt some neighbor's or friend's suggestion. The cure may be worse than the disease, as was the silly case where moth-balls were tied to each plant!

One of the most common means of combatting insects and disease is to breed new disease-resistant varieties. This is merely a palliative. They enjoy a brief respite and soon fall prey to the old diseases.

To prevent the inroads of root-knot nematode in peach orchards, plants which are immune to that pest are grown under the trees. Since these cover crops are extremely unappetizing to the pest, the latter do not multiply in sufficient quantity to hurt the trees. In the state of Georgia, the Government experimented with one such crop, a legume called *Crotalaria spectabilis.* With its use the peach crop output soared, but here again, there seems to be a hitch—it is believed that this legume is poisonous to livestock.

All of these are negative approaches. There is only one positive remedy and that is to use organic compost and to follow other cultural practices that will make the plant or tree strong enough not to fall by the wayside as soon as the wind blows from the wrong direction. We must not expect perfection. There will always be some destructive insects. There will always be some disease on a small scale. They are Nature's way of eliminating the weak and perpetuating the stronger, hardier strains. But we don't have to take measures like those of primitive surgery, which were satirized in the old vaudeville gag: "The operation was a success, but the patient died."

Are arsenate and lead compounds which remain as residues on fruits and vegetables poisonous? Read *100,000,000 Guinea Pigs* by Kallet and Schlink and you will be absolutely convinced that they do incalculable harm to the human system.

The Federal Food and Drug Administration permits fruit to be marketed that contains 12/1000 of a grain of arsenic, but in many instances cases are found where they contain double that amount. Many innocent headaches and digestive disturbances are due to taking in small quantities of these poisons. Some physicians believe that flu and grippe in some instances are due indirectly to poisonous arsenates that find their way into many foods. You have heard of many country children who get sick from eating green apples. It is not always because the apples are green, but because they eat fruit that has not had the spray washed off. Walking through country lanes, one sees apple boughs laden with fruit that has a seemingly innocuous white powdery coating, which is the spray. Many people refresh themselves as they pass by such orchards and then come the cramps.

European countries are much more strict in connection with poison spray residues on food. England does not permit any American apple to be shipped there unless it contains arsenic residue of less than 1/100 of a grain per pound.

"In 1927," say the authors of *100,000,000 Guinea Pigs,* "the Federal Food and Drug Administration declared that while apples intended for export must not bear a residue of more than 1/100 of a grain of arsenic trioxide per pound, *apples intended for domestic consumption would be permitted to carry two and one-half times the safe limit of arsenic!* Nevertheless, we have every reason to be grateful to the British authorities, for before 1926, apples were legally sent to market with any amount of arsenic that happened to remain in and on them, even if it were ten or twenty-five times the safe amount."

No matter how thoroughly the fruit is washed there still remains some residue of poisons and what is more important is the fact that some of the arsenic and lead is absorbed into the fruit and cannot be washed out. It has been found that when paris-green, another poisonous insecticide, is applied to the potato plant at night, it is found *in the potatoes* the next morning.

Many vegetable plants are badly polluted with poison sprays. This is especially true of the cabbage family, including broccoli and brussels sprouts. In eating cabbage from market, always be sure to throw away the outer leaves.

I am going to quote from an article by J. Eaton called "Backward Farmers," in the September, 1939, *Forum:*

"One season many years ago, all the spinach growing in the Netherlands was attacked by a terrible blight. To cure it the farmers were commanded by royal edict to spray the plants with a certain strong antidote. But, when the Hollanders ate the spinach, an epidemic of intestinal infection promptly followed. Then all sales of the vegetable were banned—almost all, that is. One consignment of the greens coming into the Hague had proved innocent of harm. Why? Because, having no blight, the plants had not needed spraying.

"When this fact was discovered, the whole town—market men, doctors, and citizens—focused its attention on the grower of that lot of spinach. By what magic had his garden escaped the plague? Agricultural officials were at once sent down to interview Dr. Ehrenfried Pfeiffer.

"These august visitors to the experimental farms listened in amazement to the scientist's report.

"He used no chemical fertilizer.

"How, then, could he produce such abundant crops?

"He used none of the sprays recommended for pests.

"Why, in that case, wasn't his potato patch speckled with bugs, and his fruit gnawed through by worms?

"The inspectors learned first that the rich earth in his fields was constantly renewed by means of composting all the waste matter on the place—weeds, vegetable stalks, grass cuttings, garbage. Manure was composted also, and was never used in a raw state. In a shady place stood the slanting, earth-covered rectangles of the compost heaps.

" 'Four months ago,' remarked the farmer-scientist, 'this particular pile was manure. Eight months ago, this one was table scraps, eggshells, moldy bread, parings and fruit skins.' "

6

SAFE MEANS OF CHECKING INSECTS AND DISEASE IN GARDENS

We acknowledge that there will always be some destructive insects and some disease, but if the organic method is followed they will be cut to an irreducible minimum. In our organic garden there is so little of either that we do absolutely nothing about them. There are safe measures however that can be taken by gardeners to keep the insect front under control.

Many gardeners take potato bugs off plants by hand. They readily come off with an easy flick of the hand into a pan below filled with kerosene or other strong liquid.

The earthworm should be assiduously cultivated in the vegetable garden because he eats the larvae of destructive insects. There is definitely a connection between a large earthworm population and a small amount of disease and destructive insects.

A great deal of advice is given about the growing of certain plants to drive away certain insects. You are told to plant mint near cabbage to repel the moth. The late Dr. Rudolph Steiner of the Bio-Dynamic movement suggested the planting of nasturtiums among fruit trees to reduce destruction by aphids. Followers of the Bio-Dynamic method go a step further. They paint and spray the trees with an extract of nasturtium. About this, Dr. Pfeiffer has this to say in his book, *Bio-Dynamic Farming and Gardening:*

"Such plantings are effective because the nasturtium contains a strong aromatic substance which also penetrates the ground through its roots. This the trees can take up through their roots and bring into their sap stream, making changes in it which are imperceptible to us but very evident to the fine organs of smell and taste possessed by the insects. Result: the aphids disappear."

He also states:

"Since the cabbage butterfly is repelled by hemp, tomatoes, rosemary, sage or peppermint, it would be wise to use such plants as protective intercrops. The asparagus beetle is repelled by tomatoes, the mole cricket by hemp.

"To combat the 'capsis-fly' the blossoming tree must be given a highly diluted liquid spray of slaked lime. It is often also advisable to sow or scatter the seed of 'attracting plants' to entice insects from plants needing protection. In this way we can use lettuce, spinach and potatoes to combat cockchafer larvae, strawberry weevil, wood-louse, and ear worm. There are many possibilities for helping oneself in such natural ways. For instance, in order to catch snails the empty half orange or grapefruit skins may be laid between the beds with the open side down; after a certain length of time a whole collection of these pests will be found under them."

In another section of the same work, Dr. Pfeiffer discusses tactics against flea beetles:

"These are driven away by shade. They prefer a crusty soil, the surface of which, swept by the wind, becomes dry and impermeable. Hence we should work to develop a crumbly soil with strong capillary action, and to make a shade by means of mixed crops and intercrops. We should also mulch the soil between the rows with partially rotted leaf compost. The flea beetle shuns tomatoes and wormwood. It is therefore advisable to plant these at random, here and there, between cabbage or radish plants. Even the spreading of the trimmed off shoots of tomatoes can be helpful."

While on this subject here are a few words of advice issued by the Connecticut Agricultural Experiment Station:

"Avoiding the use of ornamental plants which attract pests is one of the best ways to overcome the fungicide and insecticide shortage next season. In place of such insecticide-consumers as roses, dahlias and gladioli, we might substitute phlox, marigolds, calendulas and asters, both annual and perennial. Dahlias are troubled by the corn borer, on which nicotine is used. Gladioli are susceptible to thrips, for which the control has been rotenone or tartar emetic. Roses are heir to any number of insects and diseases which call for spraying or dusting at ten-day intervals."

To refer back to nasturtiums, a friend of mine reports that he and his family have always planted nasturtiums in among cucumbers and squashes. This, he says, discourages the squash and cucumber beetles, and the black aphids or lice that attack nasturtium do not appear, either. This trick he learned from a Pennsylvania Dutch great-aunt who always grew flowers in her vegetable garden, and never had any insect pests to speak of.

Mrs. Thomas Swanton of Webster, N. Y., writes, "My grandfather, a Canadian farmer, planted onions around his peach trees; so do I. Each summer I dig up the old ones and plant young. Borers do not like the odor."

There are any number of suggestions as to how to defeat the actions of the cut-worm. Some advise a ring of wood ashes in a circle around each plant. Soak the ashes, which shouldn't touch the plant. Cutworms usually attack beans, brussels sprouts, cabbages, cauliflower and tomato plants. Some suggest putting pieces of board as a trap here and there in a row. Cutworms like to hide underneath boards. Some people make paper collars which they put around the stem so that half of it is above ground and half underneath. These are narrow collars that hug the stem.

Wood ashes spread on the ground will discourage the activities of slugs and snails, although a certain amount of these creatures are to be welcomed as part of Nature. Some put wood ashes on cabbage plants to kill disease organisms and insects. And it is said that wood ashes on plants will drive away rabbits.

The lower leaves of lettuce are sometimes attacked by a rot which spreads and kills the heart of the plant. This can be prevented by placing clean sand under the plant. For the same reason straw mulches under plants like watermelon and muskmelons prevent rotting from contact with the ground.

Burns Byram, county agent of Tama County, Iowa, developed a chinch bug barrier which can be laid by two men at the rate of about 80 rods an hour. He uses a device that looks like a sled from which paper unrolls and is placed in a ditch made by the sled as it goes along. Baffles throw dirt against the paper as the sled moves. Devices like these do not harm the soil in the fight against insects, because no poisons are used.

A very fine machine has been invented to fight the boll weevil. The inventor is Alex R. Nisbet of Plainsview, Texas. His device consists of a big fan, connected to the front of a tractor, which sucks the weevils off the plant into sacks. Hundreds of pounds of boll weevils, bollworms, leaf worms, hoppers, etc., are destroyed from each acre. This machine works just as well on potato plants. He builds the machines himself and sells them at a price of $275. One user said, "One operation did more good than three applications of poison." Another large cotton farmer who used Nisbet's machine secured a bigger yield from 90 acres than on the rest of his crop on which he dusted poisons.

Developments like this may lead to an interesting competition between mechanical ingenuity and the chemical industry. Perhaps the small gardener will even be able to extend the family vacuum cleaner into the garden and make short shrift of damaging insects.

According to *Science News Letter* of November 7, 1942:

"Very effective control of saw flies is reported by C. S. Noble, Canadian pioneer of subsurface tillage. The grubs are raised and left in a more exposed position, he reports. Tests show that a large percentage of them do not survive.

"Grasshoppers can be controlled by leaving strips of land unbladed. These are traps. Grasshoppers lay their eggs in the nearest firm soil. Then late in the fall, and again early in the spring, the trap strips are lightly cultivated. This destroys nearly all 'hopper eggs, Mr. Noble declares."

For controlling squash bug a friend suggests that the squash be planted very early. The squash borer usually lays his eggs in July so that by early planting the vines become so large and well grown by that time that the borer can do very little damage.

With our typical American mechanical ingenuity we can produce many simple devices which will spell doom to bugs, borers, and the whole family of noxious insects. Further studies in ecology (that branch of biology which treats of plants and animals in their natural habitats and their inter-relationships) will open up many vistas of agricultural thought, and folklore will contribute to it, without doubt.

The biological method of insect control is coming into wider use. On a large Florida sugar cane farm there were recently turned loose thousands of frogs shipped in from Argentina to combat insect pests that were ruining the crop. Here is an interesting item: English gardeners make a practice of turning vast numbers of toads loose in their gardens. The average toad consumes daily—or rather nightly—harmful bugs to the amount of three times the capacity of its stomach. Because the toad is nocturnal it is especially effective against cutworms and slugs.

Pathfinder, June 26, 1944, reported:

"Now southern cotton planters are standing by while *USDA* engineers plot a war on the dreaded cotton leaf worm with the aid of countless insect eggs imported from Argentina. When hatched, a war-like breed of parasites will invade the cotton plantations in an all-out attack on the leaf worm.

"An Argentinian discovered the parasite was a lifelong enemy of the cotton foe. Later experiments by Harry Parker of the *USDA* proved the efficiency of this means of combatting the leaf worm and mass importation of the insect eggs followed."

In 1868 California citrus trees were attacked by the cottony-cushion scale which threatened to ruin the industry. Government investigators discovered that in Australia this disease was kept in check by lady-bug beetles of a certain species. Some of these were imported and within a few years they ate up practically every cottony-cushion scale in California.

Today in California mealybugs on citrus trees are being kept in check biologically by the introduction of certain kinds of lady bugs that are the natural enemy of the mealy bug. The State maintains insectaries at sixteen different locations where lady-bugs are propagated for this purpose.

Late in the summer, some gardeners range far and wide collecting from the tall grass the egg containers of the praying mantis. These they place in their gardens so that in the spring, after the egg clusters hatch, their gardens are fairly teeming with the bug destroying mantis.

Perhaps these methods may require a little extra thought and effort. There is a question whether the indiscriminate application of the spray gun is easier. But there can be no doubt as to which is the safest from the standpoint of soil and human well-being.

The gardener who has always sprayed may be skeptical. He may not readily accept our main contention that the continued use of humus as the exclusive fertilizing agent will bring such vigorous health to the plant that there will be less disease and insects. That is to be expected. The continuous propaganda of the chemical interests has left its mark, and mere words in a book will not easily erase such a scar.

But there is an intoxicating thrill that awaits such a man or woman, even in his first season of gardening in the organic manner. He is in for some surprises.

7

ORGANIC ORCHARDS

Can the use of chemical fertilizers in the orchard be abandoned, and the organic system used there exclusively? If you tell a farmer that he can raise fruit without spraying he will look at you with that queer expression that seems to question your sanity. Actual results show it can be done. However, in an established orchard the change-over will take many years.

Quality fruit was raised hundreds of years ago, long before sprays were known. Talk to old-time farmers. Not only will they tell you that chemicals were not used, but that the taste of the fruit then was more delicious. In the old days of mixed farming the soil was fertile enough to grow strong trees that could resist disease and insects.

Luther Burbank in *Partner of Nature* wrote:

> "It has long been my belief that a solution of the problem of protecting our fruit trees from both insect and fungus pests must eventually be found in the development of immunity in the trees themselves, rather than in resort to such expedients as spraying and 'gassing'. . . . In this regard we may well take a leaf from the note-book of the physician, who has learned that immunity to disease often depends more upon the condition of the patient than upon the presence or the absence of disease germs."

The fruit orchard presents the toughest problem on the farm, in the matter of changing over from the chemical to the organic system. With ordinary crops you start with seed, but with an established orchard you have old, weakened trees, and it may take two, three or more years to rejuvenate them, and this unspectacular, gradual improvement requires patience.

Let me give you a few examples of the workings of the organic method in the fruit orchard. Helen Papashvily (author of *Anything Can Happen)* made her discoveries in rather accidental fashion:

"In 1933 we bought Ertoba Farm in Pennsylvania, 76 acres of basically good but sadly abused land. On the place was the disheartening remnant of what had once been a flourishing home orchard of approximately 50 trees, peach, apple, pear, plum and cherry. Unpruned and badly nourished for years, the trees were spindly in growth and at the time we took possession (July) covered with twisted wormy fruit and festooned with ghostlike webs of tent caterpillars.

"The first year we left the orchard alone as uneconomical to handle, while considering some other projects, notably the starting of a poultry flock. To house this flock we moved a portable brooder house down under one of the pear trees, a barren Seckel pear.

"The flock consisted of approximately 200 to 250 chicks, and clean chopped straw was used for their litter. It was our practice to use a relatively shallow litter and to change it more frequently than is customary. We disposed of the straw and caught droppings by brushing them out through a trap at the back into a pit beneath the house which was protected from the chicks on range by a heavy screening around the foundation.

"At the end of the season the brooder house was moved to a new location and the matter forgotten except that the straw and dropping pile which had been decomposing in the shallow pit under the house appeared somewhat unsightly when exposed. It was turned a number of times in the months that followed without appreciably decreasing its bulk and the next year in the interest of neatness this rather crude compost heap was worked in around the base of the pear tree to a radius of approximately 6 feet. The following spring the pear tree was a bouquet of bloom in contrast to the poor showing of its neighbors. The tree set fruit early, showed little or no cast fruit (the ground around the other trees was covered early by drops) and bore a satisfactory crop of 5 bushels of well formed russet pears in the fall.

"So unwilling are we to learn or to correlate events logically that we considered this bounty was due to the fact that it was the 'good' or bearing year of this particular tree. But when this crop was repeated the second and then the third and fourth season while trees a few feet away on every side continued to have barren years indefinitely some idea of the value of organic composting and its use in orcharding began to penetrate. Sufficient to say this yield continued with some seasonal variation and in the fall of 1942, the pear tree bore its 7th consecutive crop of 6½ bushels of clean healthy satin-firm fruit that can be eaten, full of sun, directly from the tree without first being washed, brushed or de-poisoned.

"The general physical appearance of this tree is also worthy of note.

In an orchard of thin rather pale trees afflicted with blight, mildew and caterpillars, this pear tree is heavily leafed with a dark rich foliage and the bark is clean and apparently free of fungus."

Howard, in his *Testament,* describes his own experience with diseased worn-out apple trees:

"I have just obtained confirmatory results which prove how important humus is in helping a mycorrhiza-former—the apple—to throw off disease. In 1935 I began the restoration, by means of humus, of my own garden, the soil of which was completely worn-out when I acquired it in the summer of 1934. The apple trees were literally smothered with American blight, green-fly, and fruit-destroying caterpillars like the codling moth. The quality of the fruit was poor. Nothing was done to control these pests beyond the gradual building up of the humus content of the soil. In three years the parasites disappeared; the trees were transformed; the foliage and the new wood now leave nothing to be desired; the quality of the fruit is first class.

Herbert C. White of Gardena, California, wrote me in August, 1944:

"I am enclosing a small photo of my peach tree. This was treated last June with a heavy application of partly composted material from one of my heaps and then heavily mulched with straw. I did not use any poisonous spray, or insecticides. The crop was perfect, with no sign of infection of any kind. I picked over three bushels of peaches from this young tree. I am through with all commercial fertilizers and poisonous sprays."

Rev. Robert W. Jones, Pastor of the First Unitarian Church of Memphis, Tennessee, wrote in the Memphis *Press-Scimitar* of May 20, 1944, as follows:

"Sir Albert Howard claims that the use of humus in agricultural practice will reduce the bug menace (including the cotton boll weevil) to an insignificant minimum. I have had experiences which support his thesis. I have been a small mulcher from away back. A mulcher is one who covers the ground between crops and around trees and bushes with leaves, grass, paper, etc. Mulch is humus in the making. Last year, for example, I dried about 12 bushels of apples from two trees that have not been sprayed in 15 years. I don't claim that I did not have a bug in a bushel, but I can truthfully say that there were not many defective apples on the trees. For the past five years the trees have been mulched with orchard grass. No spraying, no fertilizer, no bugs, but excellent apples. I have had a similar experience here in Memphis with potatoes."

A physician writing in the *American Eagle* of Estero, Florida, in July, 1944, said:

"I was taken to a farm recently to treat a sick friend, and as we drove through various orchards, my friend, a citrus grower, began talking about citrus groves. He pointed out the various ones that were fertilized with chemicals, how polluted the leaves were. Then he showed me extinct groves that had perished by certain diseases. 'What disease' I asked him, and he candidly replied, 'stupidity. But they don't learn', he said, 'for after thirty years, they still pour on the sulphur, brimstone and fire, and they plough deep, the deeper, the better, they say. Even in the dry season with such a lack of moisture, and even though they pump water, it is still too dry. They burn up everything around the plants; not even bacteria can live, far less earthworms. Look at my grove, the leaves are clean, the ground is "dirty," sure; it is full of trash, humus, grass, leaves, all sorts of junk, but the earth is full of bacteria and earthworms. It is moist, mellow and lovely to handle. My fruit is sweet and beautiful. I don't have to spray anything.' "

On my own farm, I began, last year, to nurse along a few older trees with compost and mulches. No poison sprays are used. An improvement can be seen in this second season but it will probably take several more years for their complete rehabilitation. It is a slow process to restore an orchard, but the results are sound and worth while.

To restore an orchard to real fertility it is best first to weed the ground clean underneath the trees to about a foot or two beyond the outside drip-line of the branches. Then the soil should be given a good liming, for most fruit trees require a sweet soil. Lime is especially helpful where chemical fertilizers have been used, as the soil will in that case be acid. Whether chemicals have been used or not, give the ground a strong lime treatment. If water is readily accessible, then water heavily. This will help the lime to penetrate deeply and quickly.

The next thing to do is to spade compost in the top two or three inches of soil. Start a foot away from the trunk. There are no feeding roots there. Spread it to about a foot or two beyond the drip-line of the outer branches. A tree does not need too much fertilizer. A thin layer, less than one-half inch should be sufficient. Raw, unrotted manure should never be applied to a fruit tree. This is well emphasized in *Farmers' Bulletin No. 1522* of the U. S. Department of Agriculture which says:

> "Manure does not readily rot under Great Plains conditions and the excessive application of manure may actually injure the trees. Such results are evident in one of the cultural experiments at the Northern Great Plains Field Station, where blocks of trees that have been manured heavily over a period of years are in weaker condition than those in clean-cultivated blocks or in blocks receiving other treatments."

Of course there is all the difference in the world between raw manure and finished compost. The next step is to place a heavy mulch under the tree at least a foot or more thick. This can be made of straw, spoiled hay, ground corn cobs, grass clippings, leaves, weeds, etc. There are many reasons for this mulch but one important one is to prevent grass from growing under the tree.

Sir Albert Howard's famous controlled experiment at Pusa demonstrated in no uncertain fashion that grass was harmful to most fruit trees, serving "as an asphyxiating agent to the roots of all the fruit trees investigated except the guavas. . . ."

Soil aeration is of vital importance in agriculture. As Howard has pointed out in his classic essay on that subject, Chapter IX of *An Agricultural Testament* (which details work on orchards at Pusa):

> "The transformation of soil fertility into a crop is only possible by means of oxidation processes. The various soil organisms—bacteria and fungi in particular—as well as the active roots need a constant supply of oxygen. . . .
>
> "The full significance of soil aeration in agriculture has only been recognized by investigators during the last quarter of a century. The reason is interesting. Till recent years most of the agricultural experiment stations were situated in humid regions where the rainfall is well distributed. Rain is a saturated solution of oxygen and is very effective in supplying this gas to the soil wherever percolation is possible."

Grass, as Howard's work demonstrated, keeps rain from penetrating quickly and deeply enough around the roots of fruit trees to keep them supplied with the necessary oxygen, and that "only when the humus content is kept high can adequate permeability be maintained."

Mulches trap and store rainwater and the bottom layer of mulch is almost always wet, keeping the soil cooler in summer and warmer in Winter. The aeration of the surface soil is better; the surface roots feed in the portion of the mulch that has turned into humus. The decay of the mulch supplies many valuable fertilizing elements.

The research in this field has been rather extensive, and shows that large amounts of available potassium are found underneath old straw mulches as compared with adjacent cultivated land. In the *Journal of the American Society of Agronomy*,[1] I. W. Wander and J. H. Gourley show that where there had been a heavy mulch for 22 to 38 years in a pear orchard, upon which no potassium had been added other than that supplied by the mulch, available potassium was very high to a depth of from 24 to 32 inches, although straw contains only one per cent potassium. On the other hand, where chemical potash fertilizers have been applied for a period of ten years, very little downward movement of available potassium has been observed.

The mulching of orchards is an imitation of Nature in the forest where the falling leaves make a nice litter. It encourages the multiplication of earthworms that like the moist condition found under the mulch. This practice is becoming increasingly popular and is giving excellent results in increased yields.

As opposed to the simplicity of organic orcharding, the chemical handling of trees is a complicated procedure. There seems to be some relationship between potassium and magnesium. As the supply of the former increases the latter decreases. Where fertilizers are used that are high in potash a reduction in the available magnesium occurs which causes a magnesium deficiency blotch to develop. There seems to be some evidence also that nitrogen application reduces the potassium content of the leaves. Chemicalizing trees results in quantities of apples falling from the tree before maturity. When a tree has been

[1] Vol. 30, pages 438-9, 1938.

thoroughly strengthened with well finished compost and mulches, premature dropping is at a minimum.

Another disadvantage of the spray is that it seriously interferes with the ability of the leaves to absorb light for a few days. Dr. A. J. Heinicke, a pomologist of Cornell, in an experiment described in *Farm Research* of April, 1944, in which a lime-sulfur spray was applied to a 10-year-old Baldwin tree enclosed in a glass cage in an orchard "found that the photosynthesis of the entire tree was reduced about 49 per cent for the first 5 days after the spray application."

There seems to be only one kind of ammunition in the average orchardist's arsenal: chemicals. No matter what the trouble may be, what the problem consists of, the solution most in demand—most frequently prescribed—seems to be some mixture or other of chemicals.

In apple orchards, it is sometimes necessary to thin out the apples in order to get fruit of commercial size. In the past this has been done by hand. Now they are using chemicals for this purpose. The idea is to get a spray, copper sulphate for example, which will kill part of the bloom but will allow a certain amount of flowering so that a portion only of the possible fruit crop per tree will mature.

Bees are very important to every fruit orchard. They pollinate the flowers that produce the fruit. Unless such pollination is effected there would be no fruit. Recently in apple orchards of New York State, hand pollination of blossoms in apple trees had to be resorted to, because the bees did not seem to co-operate in their customary magnanimous manner. They stayed away in droves. Most likely because the poison spray residues were distasteful to them.

Then there is trouble with poisoned honey. In the February issue of the *American Bee Journal,* there appeared the following item, sent in by J. W. Powell of New Mexico:

"Reports from all cotton growing areas show that more bees are being poisoned than ever and from samples of pollen analyzed by several, it would seem that most of the poisoning is caused from the dust in the pollen itself. Sometimes bees die late in spring. When the bees get the poisoned nectar from the nectar glands of the cotton, they die at once but the poisoned pollen can be stored in the hive and kill the bees as they eat it. As the cotton bloom is not open at night, I believe all beekeepers in cotton growing districts should make an effort to compel cotton growers to do their poisoning at night, since this would save many bees and it would be better for the farmer as the dust sticks to the leaves of the cotton better at this time because the leaves are wet with dew and hold it better.

"The newspapers have made such a fuss over poison being found in the pollen that there are many people who will not buy cotton honey, although the chemists say that there is not enough to be obtained in that way to hurt anyone. I do not know the weight of dust used for poisoning compared to honey and yet it is my opinion that this dust would rise to the top of honey and could be skimmed off, even if there were enough to cause trouble. However, we know we are losing bees from this cause and I think it is the duty of beekeepers to get assistance in this problem."

Draw your own conclusions! Bees are extremely important to farmers. Besides pollinating fruit trees they are needed to pollinize farm crops such as alfalfa, clover, vegetables, etc. Due to the great use of poison sprays on many cultivated crops which bees are bound to consume there is great danger not only to the health of the bees but also that such poisons will show up in the honey. Beekeepers' problems have increased many times over in the last few decades—since spraying became standard practice. Experiment stations are beginning to be concerned over the multitudinous new diseases of the bee.

Bees located near cider mills sometimes get diarrhea from feeding on apple pomace. Apples usually are heavily sprayed with poisons. Watch your cider. Is *it* free from poison spray residues?

The all-important question is: Can a fruit orchard that must produce an income each year, quickly change over to the organic method? The answer, unfortunately, is "no." In such a complete change-over there would be a transition period of probably low yields which might throw the relationship between debit and credit on the ledger out of kilter. Therefore, the orchardist who is desirous of throwing chemicals overboard must adopt some kind of plan; let us say, a several-year-plan. It might be three years or ten years. The age and condition of the trees and the status of soil fertility will determine. A rotation plan can be worked out, with each year a certain portion of the orchard added to the organic section—like the fallowing plan of good farmers from antiquity to the present.

Fruit naturally grown on healthy trees flourishing in fertile soil, largely insect and pest-free, is superior in taste, color and health-giving properties, and can command a premium in the market.

Recently a certain company has been offering premiums of 50c a bushel for organically-grown grains such as wheat, corn and buckwheat. It has done this because it can sell them at high prices to discriminating purchasers. Here is a tremendous field waiting to be tapped; a large part of our population is not only health-conscious but has the money to pay a little more for quality food that will keep the doctor away. The same holds true of fruit. The time is not far off when grocers will offer special grades of organically-grown fruit, guaranteed to be unsprayed, at higher than regular prices and they will have difficulty meeting the demand.

PART V

Good and Bad Farming Practices

I
SOME PRACTICE—GOOD AND BAD

As stated in previous chapters, the making of compost is not the whole program of organic farming or gardening. There are other cultural practices that must be observed to get the land into good heart and to produce healthy plants and farm animals.

For example, crops must be rotated. Some weeds follow certain crops and unless there is rotation, these weeds will gain an ineradicable foothold. Then there is the fact that different crops make different demands on the nutrient supply of the soil. Clover, for example, may absorb more calcium than barley, while the latter uses up much more silicon. Disease organisms and pests follow certain crops and unless there is a change, they tend to gain a permanent foothold.

Certain crops like corn, potatoes and tobacco are heavy feeders and must not be grown too often on the same soil. A rotation may be figured out of four, five, or six years. Sometimes as high as an eight-year rotation is used in which pasture figures for three or four years. There should be an intervening leguminous crop such as alfalfa and clover to furnish nitrogen. Farmers years ago did not always understand the value of and reasons for rotation. Sometimes they would raise wheat from year to year until they couldn't get a fair crop. Then they would switch to corn until it too protested. They would then go to barley and finally to beans—a practice anything but sound.

Is it any wonder that we hear of blights and plant plagues that occurred hundreds of years ago? The opponents of the organic method usually bring this up and ask why there should have been such plagues then, when no chemical fertilizers were employed. The answer is, wrong cultural practices, lazy husbandry, mismanagement of manure and general ignorance. Mining the soil rather than husbanding it.

In the vegetable garden where unusually large quantities of compost are spaded in each year the question of rotation is not anywhere near as important as on the farm, but on general principles, plant locations should be changed from year to year.

It is important to know which crops demand an alkaline condition and which an acid one. Where compost is made specifically for an acid-loving crop lime should be omitted. Vegetables that do not need lime are potatoes, parsley, radishes, turnips, squash and watermelon. It is rather strange that muskmelon actually requires large amounts of lime, whereas watermelon prefers an acid soil. Other plants that require an acid soil are strawberries and the other berries.

Where land has been abused for a long time and its sub-soil becomes extremely hard-packed, a subsoiler should be used to open it up. Where compost is used liberally over a period of two or three years, subsoiling will occur naturally, especially with the aid of earthworms which will be encouraged to multiply. In the same way hard-pans will be broken up and tilth restored.

In planting potatoes we have found that it is closer to Nature to plant the whole potato rather than cutting to a single eye and planting the pieces. Nature plants whole seed potatoes. When you cut a potato for seed, the cells are damaged and a protector callus forms to heal the wound; at times this does not happen and the piece rots, parasites get into it and devour it before it can be sprouted. Furthermore, in cutting to a single eye the seed piece is usually too small to furnish food to a vigorous young plant. Cutting to two eyes is better, but planting whole small seed, slightly immature, is preferable. In a comparative experiment I found that using whole potatoes gave a yield three times the weight of the potatoes grown the ordinary way. Another advantage is that the labor per pound of potato is much less in harvesting due to the greater yield per plant. This is important to the small gardener who has to harvest with hand tools.

It is our practice after growing corn, to flatten the stalks as soon as possible. We go over the field with a culti-packer or roller. Many farmers let stalks stand all winter. Where corn stubble is flattened, snow and rain saturate it. Snow, "poor man's manure," contains nitrogen and other valuable elements, and the flattened stalks serve to collect snow. Flattened stalks not only slow down spring floods but also remove earth that is contained in the moving water. Since corn crop residue is extremely heavy it is liable, when plowed under, to interfere with the next crop, so we always fallow that field, discing the ground and planting a green manure crop of legumes. We spread manure as early as possible as a sheet compost and in July or August plow everything under. In September we then plant wheat or barley.

Fallowing should be observed by every farmer. We learn from the Old Testament that the Hebrew farmer fallowed every field at least once in seven years. It was part of his religion and he was considered an outcast if he did not do it. It rejuvenated the soil. Irish farmers fallow their land every seventh year, also, and those near the sea mulch it with kelp—sea weed. There are several methods of fallowing. Ordinarily the land is just left idle. However, it is advisable to go over it with a mower about three times during the summer to cut the weeds down. Aside from preventing weeds from seeding, this covers the land with a mulch which is very valuable in catching rain and making humus. Finally, fallowing conserves moisture where rainfall is light.

In looking over a field of oats recently, I noticed a long strip down the middle where the crop was higher than the rest of the field. This excited my interest and I was told that at that place there had been an old hedgerow dividing two fields. They had cut it down to make one big field of it, in order to save labor in farming operations. The hedgerow represented a fallow of many years standing, the weeds and leaves enriching the soil, year in and year out.

Do you remember, during the depression years when tremendous surpluses were produced on our farms which our warehouses could not hold? Governmental subsidies were paid for plowing under crops. Such days may return again. Modern machinery has speeded up farm production and future developments in the machinery line may accentuate this problem. Would it not be smarter to urge that a certain portion of the arable acreage of each farm be fallowed each year and pay subsidies for that?

Mulching is a practice that should be used by every gardener. The mulch simulates Nature. Soil bacteria are stimulated under mulch and a mulched crop has the benefit of a vastly greater bacterial population. Moist earth under mulch encourages the multiplication of earthworms. Mulching prevents the incrustation of the soil surface by the pounding effect of the rain which might hard-pack the soil surface.

At least one tomato grower I know of has recently doubled his crop by mulching. It is excellent for potatoes also and should be started as soon as they are planted. The mulch can be applied to the entire ground and the plants will grow right through it. Many growers mulch acres of corn. Such plants as pumpkins, squash and melons should be mulched so that they will not rot by touching the ground.

A mulch should be spaded into the soil the next year unless it is too thick. In the latter case it can be used again. Where peat-moss is used as a mulch it may be spaded in for two or three seasons only. If continued after that, it may dry out the soil. In such case it should be raked aside and used again as a mulch.

It has been mentioned before that manure from animals that are fed organically-grown food will decay much faster than other manures. This applies also to green matter. If you have raised a cover crop organically, or if you plow under organically-raised straw, it will usually decay without interfering with the next crop.

Green manuring under ordinary conditions may end in disaster if the season is unusually dry. Bacteria use up a great deal of moisture. Where soil is full of humus it stores more water available for the breakdown of a green manure crop.

I like the little story one of our farm-hands told me. He was once applying for a job on a farm and was asked "Do you ever talk to the chickens?" He applied in the affirmative. "That's good," said the farmer. "I never hire anyone unless he does. It shows he will take good care of the chickens." This "good care" attitude is all important—not only for chickens but for other farm stock as well.

An absolute violation of good farming is the unbalanced type of agriculture, which goes into various forms of single-crop specialization, such as cotton, corn, or tobacco on large acreage. Typically, such farmers have no livestock to furnish manure and depend entirely on chemical fertilizers purchased by the carload.

There are smaller farmers who grow wheat, corn, oats or other single crops, but buy their milk, butter, eggs and vegetables at the village grocery. Such farm operators are squandering the soil's capital. A farm should be a well-balanced agricultural unit and should always have livestock to furnish manure. Farms should not be too large. I strongly feel that legislation should be enacted preventing the ownership of more than a certain number of acres varying with the location and productivity of the soil. Vast-acred, assembly-line, single-crop farms ought to be outlawed, or strictly controlled for fertility maintenance.

A question often asked is: "Should the organic farmer use modern machinery, or should he be a 'true son of Nature' and do all farming operations in the old-fashioned manner?" I think we must use common sense. The world is advancing. Wonderful machinery is being developed. We don't have to be hermits, out of adjustment with modern society. It is nice to use horses but sometimes it is necessary to have a tractor. Many farmers have both. Unfortunately no tractor has been placed on the market yet that gives manure. There is no question, however, that the tractor is expediting the mining and exhaustion of our soil, but this can be counteracted by a return to organic farming practices.

A machine which is definitely not to be recommended is the combine, which cuts and threshes wheat in one operation without having the grain stand in the field to season. For centuries wheat, barley and other grains have been cut down at harvest and allowed to stand in shocks in the field. The grain continues to ripen in the shock. The old-fashioned farmer didn't thresh his wheat even after it stood in the shocks but stored it in the barn, threshing it as needed. There is a further ripening in the barn. With the combine the wheat is taken right off the field immediately. The ripening process stops abruptly. Due to the abrupt stoppage of the seasoning of the grain with these big combines the flour mills have to treat it with chemicals to make it "keep." It goes bad much more quickly than the old-fashioned fully matured kind.

An increasing amount of our food is grown in greenhouses. Some of them are so large that plowing is done by horses and by tractor. In a house of this kind one acre can yield over ten times the amount of an outdoor acre, and the conditions are artificial. Four crops are grown in one year on the same soil. Tremendous amounts of chemical fertilizers are used. There are pipes three feet below the surface which spray steam to sterilize the soil to kill dangerous organisms.

Then we have soilless culture which I believe is far worse than green house gardening, because it is even more artificial. Some of it is done in tanks, some in gravel without benefit of soil microbes or the rare elements in the soil the presence of which we are unable to detect. Corporations are already going into this kind of production on a commercial scale and vegetables such as tomatoes and potatoes are appearing in the markets grown by this method, but they are not so labeled.

The liberal use of humus where irrigation is practiced is extremely important, in fact more so than under ordinary cultural conditions. Sir Albert Howard in the *Testament* says of this practice:

> "The water must be used in such a manner and the soil management must be such that the fertility of the soil is maintained intact. There is obviously no point in creating, at vast expense, a Canal Colony and producing crops for a generation or two, followed by a desert of alkali land. Such an achievement merely provides another example of agricultural banditry. It must always be remembered that the ancient irrigators never developed any efficient method of perennial irrigation, but were content with the basin system, a device by which irrigation and soil aeration can be combined. Under the basin system," says Howard, "the land is embanked: watered once: when dry enough it is cultivated and sown. In this way water can be provided without any interference with soil aeration."

Farm tenancy is responsible for many evils in agriculture. In 1880 25% of the farms in this country were operated by tenants. By 1935 this percentage had increased to 42.1. The tendency in this direction is again increasing alarmingly now that the F.S.A. is out of the picture. The average tenant is here today and gone tomorrow. He doesn't build up his land. Many tenants actually sell their manure. They violate all the rules of good farming; they plant too many acres in open cultivated crops such as corn and potatoes which permit destructive erosion during rains; they do not bother much with strip-cropping, terracing and other devices to prevent soil erosion, such as dams, storage reservoirs and contour tillage. They will plant a cultivated crop such as corn on a steep hillside, leaving it wide open to washing and to gullying. Why should they do all this? The land is not theirs!

The owner wants income only and is not there to guide or command. A wise lease should provide that no manure be sold off the farm. It should also regulate the amount of open type, cultivated crops that may be grown. In fact where mortgages are given to owners of farms, the mortgagor should write these provisions for the maintenance of soil fertility in the mortgage.

The tendency towards very small families cannot, I suppose, be called bad farming practice: it is a sociological phenomenon; but none the less, farms suffer from too few hands.

An interesting letter has come to me from a friend of mine in northern New England:

> "The French Canadians are taking over northern New England. They buy a worn-out farm and by thrift and hard work build it up to make a living for their rapidly increasing families and to pay for the farm and improvements. When a boy is 12 years old, he is an asset on a farm; he can drive a team, milk, tend the pigs and chickens, etc. Families of 10 to 12 are quite common—some run as high as 18 children. Each of them spread out and repeat the process and in a generation or two they will possess this north country. They are tough, good workers, good farmers, honest, and take pride in their farms and in their families. Too many of the native New Englanders love trips to town and the movies, some mechanical type of farming that does not take too much time; and their wives want clothes made in New York, hair-do's, the movies, and don't want to be bothered with babies."

On a recent trip to Quebec, I noticed from the train window that as we approached the Canadian border practically every farm had compost heaps. This was even more noticeable in Canada. The inference is logical and clear: with natural manures and composts, the yield of the land is richer and more life-giving, and will support a larger number of people. Our farming ancestors, typically, had large families.

The destruction of woodland and forests seriously affects neighboring lands. Wherever there are woods there will always be a higher water-table. Destroy the forest and the water level in the sub-soil goes down to the detriment of all land for considerable distances. Draining swamps has the same effect. A forest should never be completely cut. The farmer who has woodland should practice selective cutting. This gives more light and air to the trees that are left standing and makes a permanent wood crop possible.

More fertile land has been destroyed by erosion since 1914 than in the whole history of mankind. Sixty-one percent of our land in the United States (253 million acres) by 1937 had either been completely exhausted or had lost most of its fertility. Doctor H. H. Bennett, Chief of the U. S. Soil Conservation Service recently said that we have destroyed by erosion *more land in less time* than any other nation. If we continue at our present rate America will be a desert in a hundred years.

Anyone can see that soil erosion is retarded by the many organic practices described in this book. On the other hand, rain rolls off the slopes of hard-surfaced fields where chemicals have been used, gathering momentum as it goes, very little of it seeping into the ground. Large quantities of soil are washed along and eventually find themselves in the sea. More soil is blown away by the wind. The presence of decaying organic matter may be likened to tiny sponges which absorb much of this rain. Studies have been made at many agricultural colleges regarding the run-off rates of rain water on various kinds of soil. Where liberal amounts of compost have been used the run-off rate has always been found to be very low.

In *Miscellaneous Publication 394* of the U. S. Department of Agriculture it was shown that "the run-off from grassland (pasture) during a 6-year experimental period carried with it an average of only 0.09 ton of soil per acre annually . . . a plot planted to corn continuously showed an average annual soil loss of approximately 112 tons per acre during this same 6-year period. That amounts to more than two-thirds of an inch of soil each year."

Organic practices help to create a soil structure that will absorb a maximum amount of rain, prevent erosion and provide more moisture for growing crops.

2
TO PLOW OR NOT TO PLOW

A book published recently by the University of Oklahoma Press, *Plowman's Folly* by Edward H. Faulkner, denounces the use of the mouldboard plow. This book has created something of a sensation in the public prints. Practically every magazine and newspaper has discussed it, mostly in a favorable light. "Down with the villainous plow—the curse of modern agriculture," they say. Even some of the agricultural college professors have come to lean strongly toward Faulkner's teachings regarding the plow.

Plowman's Folly advocates discing and harrowing the land instead of plowing; a theory and practice I consider ill-advised. A great many persons interested in the organic method of farming seem eager to embrace his theory, perhaps in their subconscious mind associating the plow with the high-production and artificial methods of present-day commercial farming. They should bear in mind, however, that while this useful instrument might cause trouble under certain conditions, it is a safe and efficient tool where compost is used in the soil as the fertilizer element, and in other situations as well.

There is no question that in some sections of the country, for special reasons, plowing can be dangerous. Take the case of the Great Plains region, where so-called "dry-land farming" is practiced. Because of dry sandy soil and skimpy top soil, and an extremely low annual rainfall, plowing would be an open invitation to the wholesale blowing away of the soil. Special tools, therefore, have been designed for farming there. There is a lister, which has V-shaped sweeps which are drawn underground four to six inches. The ground underneath is rippled while the surface is not disturbed. If organic farming in these dry-land sections were practiced and the soil enriched with compost instead of chemicals there possibly would not be any need for special tools or a special kind of farming.

I have one of these listers. In an experimental trial our tractor couldn't begin to pull it, our soil being clayey and compacted following a decade of bad farming practices by the previous owners. A neighbor who had a heavier tractor came to our aid. Still the lister wouldn't pull successfully. We are keeping this machine and will try it again in a few years when our soil has become softer with the yearly addition of compost.

There are other sections of the country, parts of the south, for example, where the structure and condition of the soil is such that it is not safe to plow. In our discussion we are not taking into consideration these "problem" areas, but are dealing with the normal or average.

I knew of Faulkner's theory through articles in *Country Life* and our own *Organic Gardening* and decided to try his method on our farm. Of a six-acre tract planted to winter wheat, we plowed one-half, while only discing and harrowing the other portion. Incidentally, because of the extra discing, it required a good deal more labor in the "no-plow" portion to get the weeds worked into the soil.

The results were quite negative where no plowing was done and weeds had a field day. Where land was plowed, the fresh weed seeds were "done" under, and by the time they worked their way up the wheat was way ahead of them. In the "no-plow" section the weeds came up faster than the crop, the result being a very poor stand compared to the other three-acre section.

Another reason why there was no thick stand of wheat was that the drill machine that planted the seed was badly hampered in its work by the stubble and plant residues on the surface which were not plowed under. I can't conceive of a machine being perfected that can give a correct coverage of seed when there is so much trash-covered sod on the surface. Imagine a previous crop of corn with thousands of stalk stumps per acre and try to visualize laying down seed for wheat, oats or any other close-growing crop. Mr. Faulkner himself admits that you could not disc down the sod of such previous crops as alfalfa and other grass pastures. These have to be plowed under.

We tried the "no-plow" method in our 1943 vegetable garden with devastating results as far as weeds were concerned. They came up with a vengeance and you could hardly see the plants themselves.

The weeding labor was all out of proportion to the economic value of the crop.

Every experienced gardener knows that it is impossible to grow straight carrots or parsnips without deep spading. The vegetable gardener will come to grief unless he plows or accomplishes its equivalent by a good spading.

Faulkner asked many farmers why they plowed. He says they didn't seem to know, except that it was to get rid of weeds. Well, that was a rather potent reason. Faulkner argues against himself, and shows the great value of the plow for keeping weeds in check. In talking about the invention of the mouldboard plow by Jethro Tull, he says: "Their land was foul with perennial weeds which were always in the way of their farming, and would grow higher than their planted crops unless they were kept in check. With the mouldboard they could now avenge themselves on these enemies, and could add other acres to what they had been able to till before."

Nowhere in his writings does Faulkner mention the menace of the tractor which, by its own weight, compacts the soil. That seems to me a greater wrong than plowing could ever possibly be except in such "problem areas" as I have noted. The ideal condition is a loose, aerated, porous earth. I have followed the plow and watched the soil lift up and then fall. If this doesn't loosen and aerate the soil I don't know what will. Where you do not plow and in its place simply disc a few extra times, you are weighing down the land *that* much more with your tractor.

Regarding this point, Professor Charles E. Kellogg, in his book *The Soils that Support Us,* says:

> "On many soils, especially those inclined to pack into hard masses, the use of heavy tractors over the land has added to the injury of the structure. . . . With the reduction in rates for electric power, many farm operations may be done in the future with electric power located at the margins of the fields and with machinery pulled with cables in order to avoid the use of heavy machinery on the soil."

Another fact that throws a shadow on the no-plow theory is the experience of certain Wisconsin farmers. A recent annual report of the University of Wisconsin Agricultural Experiment Station stated that corn yields were reduced by 1/3 or more when a farmer merely disced down a rye cover crop before planting corn. It recommended plowing under cover-crops.

The U. S. Department of Agriculture has been experimenting with the elimination of plowing in the planting of corn. The tests have so far consistently resulted in reduced yields.

Mr. H. E. Middleton, Acting Chief of Research of the U. S. Department of Agriculture says in an article in *The Land,* issue of summer 1943, "In 1940 Mr. Faulkner had over half of his test field in corn and the crop was a total failure. To my mind, this was the most significant result of his test. However, I can find no reference to it in his book."

One of the main contentions of Faulkner is that when green matter of the surface is plowed under, it destroys the capillary action of soil water. Such organic matter, he says, absorbs much water from below as well as above it, thus acting as a barrier, preventing the water from below from reaching the surface by the usual capillary action of the soil.

The Jefferson County Farm Bureau News in an editorial directed at Mr. Faulkner, states among other things, "Faulkner says that in plowing, the residue turned under always lies as a mat at the bottom of the furrow, thus definitely and effectively insulating the furrow-slice from its subsoil. Any good plowman would object most strenuously to such a premise."

Professor Kellogg, in his book mentioned above, comments on this point: "In good plowing the soil is not turned bottom-side-up, but a little less than completely over. . . . In plowing, it is better to have the organic matter somewhat mixed with the whole surface soil." Harrowing or discing after plowing helps do this.

It is the usual custom in the northeastern section of the United States to plow 6 to 8 inches on the average, so that the stubble is pretty close to the surface anyhow. It will decay much quicker and furnish fertilizer value to the growing crop, if it *is* a few inches under the surface rather than on the surface.

Regarding the question of capillarity, soil research indicates that the amount of water moved to the soil surface by capillarity from depths below 15 or 20 inches is negligible during the growing period. Faulkner himself admits that an acquaintance of his, a crop specialist with a 20-year background of experience said, "Tests show that there isn't as much capillary movement in the soil as we believed existed—it's really unimportant in many cases."

F. L. Duley, Senior Soil Conservationist at Lincoln, Nebraska, in *The Land,* arraigns Faulkner's book in no mild manner. Regarding the question of capillarity he presents a rather amusing story. Mr. Duley

> ". . . recalls to mind an old time Farmers' Institute meeting back in my home state of Missouri. The institute lecturer was painstakingly explaining how frequent cultivation of corn would bring up capillary water from below and thus rescue the crop from drought. After he had gone on at great length and apparently thought he had convinced everyone of his hearers as to the scientific soundness of the capillary idea, an old Ozark farmer put up his hand, raised slightly out of his seat, and said, 'Eh, Puf-fesser! How much do ye think ye'd have to cultivate tu make 'er too wet to plow?' "

Even if capillarity were the factor Faulkner claims it to be, the soil of a farm that was run on the organic basis would have a distinct advantage. The biologic life of such a soil would start into action. The beneficent soil organisms would begin at once to restore normal conditions, right after plowing. Bacteria, fungi and algae, present in tremendous quantities where the soil is rich in organic matter, would break down the green matter that had been plowed under. Here is where earthworms too play an important role. They bore thousands of holes per acre right up to the top, going right through the green matter to eject their castings on the surface. These burrows draw water from below and permit rains from above to seep down into the earth. Where there is a pronounced biologic life in the soil you will find a fine flexibility, a softness, an ability to keep things going under adverse circumstances.

Faulkner suggests that the residues on the surface will absorb much rain water which otherwise would run-off and cause erosion. This is one of the most important advantages also of organic farming. Where the soil is rich in humus, rain water is almost completely absorbed. On such land you never see water puddles or muddy streams.

It is not the plow that is running down our soils. Try to tell the Pennsylvania Dutch farmer of the Lancaster area not to plow. These famous farmers have plowed for two hundred and fifty years and not on the contour either. Their farms today are in such wonderful condition of fertility that they command fabulous prices, even though their principle crop is tobacco which "pulls" hard on the soil. What is the magic they employ? Nothing more than that they have always engaged in mixed farming with a strong accent on livestock, and its by-product manure. It is only recently that an occasional tractor may be seen on their farms. I have noticed also that they get all the manure at the Lancaster stock-yards and put thousands of tons of it on their farms each year.

Another item which seems to be the direct antithesis of Faulkner's theory might be mentioned here. It is subsoiling. F. Sykes, a well-known farmer in England who has followed Howard's organic methods for years on a thousand acres, and who has produced spectacular results, suggests plowing twice instead of once, provided you sub-soil the land. In an article in *Organic Gardening,* he wrote:

"My knowledge of farming had told me two things—that you cannot farm without the plough, and that its full value can only be obtained if its operations are preceded by those of the subsoiler. To prove this one of many such stories will suffice. The subsoiler broke down after only part of a field had been done—no time to repair it, so ploughing and re-seeding with barley followed. The few acres that had been subsoiled grew a splendid crop, the rest of the field practically nothing. The next year the whole field was fallowed and subsoiled, and wheat was sown. The crop is about to be harvested and is one of the best in the county. Its yield is estimated at 56 bushels to the acre. Subsoiling accounted for this improvement.

"The details of farming technique which I have devised are important:

1. Subsoiling 2 feet deep.
2. Several ploughings.
3. Several harrowings.
4. Thorough land cleaning.
5. Two heavy-roller rollings.

The subsoiling 2 feet deep aerates the subsoil and splits into fragments the underground colloidal pan, thus opening up an unlimited store of mineral plant food which the deep-rooting varieties will seek in a very short time. The ploughings and other cultivations aerate the top five inches of topsoil and work down a fine tilth, which is indispensable to a good grass establishment."

I met a farmer at our frozen food locker who farms over 300 acres. He brought the subject up himself:

"I hear there is a fellow who is against plowing. That won't do here in the Lehigh Valley. Our soil is very clayey and the soft part of the soil extends down only about six inches. The disc won't loosen the soil below that level. I have found that you must go deeper than that with your plow and loosen some of the hard clay underneath otherwise the rains will not penetrate further than the top six inches and will wash off your land. I have seen it happen."

There is a disc plow, which piles up the earth at an angle rather than turning it over as much as the plow does. This is a compromise well worth considering. Many improvements on the plow for special use will no doubt be encouraged due to the publicity Mr. Faulkner's book has given this whole question.

The best solution of all is to follow the organic method. Get as much humus in the soil as you can year after year. Let the bacteria and other beneficent biologic organisms thoroughly saturate your land. Gradually and by natural means, increase the extent of your topsoil. Work out long rotations—including hay crops—and plow, "turning over the brown earth in rich sculptured waves." Farming then tends to be foolproof.

3

BURNING OVER LAND

There is an old saying that "he who plays with fire may expect to get burned." This is especially true in agriculture. In the spring we see much burning over of fields in order to destroy previous crop residues, but a "scorched earth" policy is better left to the military.

There are five important reasons why a fire should never be made where a crop is to grow:

1. It consumes valuable organic materials—especially humus in the upper few inches of soil.
2. It cremates earthworms and kills bacteria, fungi, protozoa, algae and other beneficial soil organisms, as well as wildlife.
3. It dries out soil moisture and helps lower the water table.
4. It destroys soil fertility by breaking down earth particles. Porosity suffers.
5. It is sometimes destructive to general property, and may, if it gets out of hand, destroy forests and towns.

In spite of these overwhelming reasons the sight of fields deliberately set on fire is all too common.

Burning off weeds is a negative form of action which must be charged off on the wrong side of the ledger. Often these fires are made by the very persons who state that running farms with the use of organic fertilizers exclusively is impractical because there isn't sufficient organic material to go around. The amounts of such material destroyed in this manner each year are tremendous and would contribute a great deal to soil fertility if composted and re-applied to the land.

In many cases the weed-growth that is burned is light enough to be plowed under easily. In such cases a better way would be to cut the weeds down and haul them in for composting. On most farms, machinery is available for this. Where weeds have been growing for many years and are very woody it is a more difficult problem. What I would do then would be to go over the field in the fall with a heavy roller, flatten it out as much as possible and then disc. The effect of the winter snows will be to soften and start decay. The growth can then be plowed under in the spring and the land fallowed for a year. This will permit the matter plowed under to completely decay before a crop is planted.

Humus, which takes much effort to build up in the soil, unfortunately cannot withstand fire. It burns quite readily. That is why peat is used as a fuel in many countries. Since in many places the humus is found only in a few inches of topsoil, it is apt to be destroyed by the heat of a fire running over the surface of the land.

Fire also kills off earthworms, bacteria and other beneficial soil organisms. The question is how far down does the effect of fire reach? I don't know, but it is a definite fact that the biologic life usually exists only in the upper four or five inches of soil in any quantity. Below that, the numbers are practically negligible. Therefore, even if the effect of the flame reaches down but one inch, the amount of destruction to beneficent microscopic organisms must be enormous.

Fire robs the soil of considerable moisture. This is a simple principle of physics which any one will admit. Heat causes water to evaporate. For that reason you wouldn't make a fire next to a compost heap. In a year of unusual drouth, the fact that the land had been burned over in the spring, might be the determining factor in a crop failure. In regions of low rainfall it would be sheer folly to burn anything on the soil. In such sections the land is often fallowed every second year, because a fallow is known to save up water in the soil and the crop gets the benefit of two years of rainfull. The application of fire to the soil would appear to be a violation of one of the simplest principles of agriculture.

J. Hunt Holley, writing in the *Farmer's Weekly* of Bloemfontein, South Africa, tells us that burning-over has destroyed the fertility of millions of acres of land in Africa. This practice seems quite common in that part of the world. Holley wrote:

"During the past seven years a neighboring farmer has had 54 floods. We had none, despite the fact that we have from 100 to 300 per cent more fall in our streams. In the heaviest rain no stream on our farm rose more than six inches. The cause of this improvement lies in humus. I have not burned vegetation since 1902, though I did in 1923 singe one farm after rain. This leaves a lot of humus for the enrichment of the soil. Humus had restored the soil and enabled me to produce a wide range of crops."

Thousands of tons of leaves are burned each year in this enlightened country, but it is encouraging to note that where the matter is explained properly, much of this unnecessary destruction is stopped.

The amount of property that is destroyed by some of these weed-burning fires getting out of hand must run into a considerable sum each year. There was a case recently where such a blaze destroyed valuable young fruit trees. Some time ago the town of Bandon, Oregon, was completely razed when a few fires started by farmers burned their way into town due to an unexpected wind coming up. Eleven people died. There was a $6,000,000 fire at Berkeley, California, many years ago which was an outgrowth of a small weed-burning fest. Six hundreds buildings were destroyed.

One might imagine that the intense heat of the fire would "consume" or oxidize the arsenic and similar metallic poisons that have been sprayed on the land and in that way at least partly pay for the damage done. But it is a well-known fact that fire has little effect on arsenic. In fact, arsenic in any form retains its poisonous qualities.

It is exceedingly dangerous to burn off grazing land in the early part of spring, as it may cause fire injury to grass roots and seed crowns. When a farmer is depending on a certain amount of hay from his pasture for summer feed purposes he may be terribly disappointed in his supply if he burns off the land cover.

In spite of all these facts some agricultural text-books actually advise the burning of fields which are badly infested with weeds. Occasionally you hear a dissenting voice, such as T. J. Murray who in a December, 1921 issue of *Soil Science* said, "The practice of burning wheat stubble on the farm is common but such procedure is detrimental to permanent agriculture."

An experience we had on our own farm convinced me of the obvious danger of making bonfires on the land. Our farm-hand made a series of small fires along the edge of a field last summer to burn up some corn-stalks, not being aware of our policy against fires. Later, the field was sowed with barley. In due time the sprouts began to make their way to the surface and the field became a beautiful green expanse *except* at those spots where the fires had been. These showed up a yellowish color. It was so distinctive that visitors remarked about it and called attention to the fact that the barley that did come through there quickly turned a yellowish color and didn't thrive as well as the rest of the field.

The 1938 U. S. Agricultural Year Book (p. 741) gives a gloomy picture of the burning of forest and its devastating effect on the soil. Some excerpts follow:

"When cut-over land is burned the fire accomplishes in minutes the degree of litter removal that would be achieved naturally only after several years. This sudden removal of the litter and its living population sets in motion a chain of events leading directly to deterioration or loss of the soil. The highly alkaline ash is leached or carried into the soil with the first rains and there acts to deflocculate the soil granules made up of silt and clay particles. The dispersed soil particles, readily taken into suspension by falling rain, are carried into the soil channels and clog them, and surface run-off and erosion begin immediately. Loss of the surface soil, reduced infiltration of water, and desiccation rapidly reduce the soil populations. At the same time increased soil temperatures result in rapid oxidation of humus, still further reducing the fertility already impaired by loss of nitrogen during the fire.

"On light sandy soils of low gradient the processes are similar except that erosion is not a factor. In its stead the essential bases are leached deeply into the soil and in effect are lost. The organic matter, the soil and little organisms, and the nitrogen disappear just as completely as though washed away, and the surface layers are left in a sterile condition unsuited to the establishment of a new forest. Outstanding examples of the results are found in the pine plains of the Lakes States, the Cape Cod section, and the pine barrens of New Jersey.

"Even though not accompanied by cutting, forest fires are a major factor in the degeneration and destruction of forest soils. They vary in intensity from the spectacular holocausts that destroy the forest completely to the innocuous-appearing leaf fire that consumes only part of the litter. The seriousness of their effects on the soil varies accordingly.

"Extreme soil temperatures frequently develop during the great conflagrations and humus in the upper horizons is oxidized immediately.

The mineral soil has much the same appearance as samples ignited in a furnace. Accumulations of heavy debris burn with such intensity that the soil is sterilized for years. Examples have been observed where tree reproduction has failed to become established on such spots more than 10 years after the fire.

"Following the great conflagrations of 1910 and 1919 in the forests of the northern Rocky Mountains, observers have reported the loss of as much as 5 inches of topsoil over extensive areas.

"The degree to which soils become compacted following forest fires is shown by the results of an experiment conducted in hardwood stands in the Central States. Water was applied to both burned-over and unburned soils and the rate of absorption per second was measured. Undisturbed soil absorbed the first application three to four times as rapidly as the burned soil. The fourth successive application was absorbed by the unburned soil 6 to 10 times as fast as by the burned."

It is usually recommended that the residues of a diseased crop be burned for fear of "contaminating" the soil and subsequent crops. We have on earlier pages demonstrated that the careful composting of such green matter makes it absolutely safe for use on the land, the heat generated in the compost heap being high enough to kill the larvae and eggs of insects.

4

ORGANIC vs. "ARTIFICIAL" POULTRY

If you don't raise your own chickens, you have probably—like hundreds of others—looked over the rather sad array of plucked birds at the butcher's in a usually frustrated attempt to get one that looked like the healthy-colored, fat and tempting home-grown chickens of your youth in the village or on the farm. And the eggs you get—how often have you remarked that they don't seem to taste as good as they used to, and have put it down to jaded taste, or faulty romantic memory?

The fact is that in the past quarter century or so, chicken raising has become Big Business. In most instances it is no longer a side-show conducted by the farmers' wives and children for extra pocket money —it has become an assembly line sort of production.

Earlier in this book, I quote Dr. Alexis Carrel's statement concerning the poor nutritional quality of the chickens and eggs one gets in market today—chickens devitalized through selective breeding for egg production, and treated, by selective feeding, as egg machines, producing eggs vastly inferior to the "old-fashioned" kind they outwardly resemble. Does this seem fantastic?

Well, let us see what has happened in the past half century. By a system of gradual selection of the best layers for breeding purposes there have been evolved breeds which lay more than 300 eggs a year. You can imagine the effect on the organs of a chicken which has been turned into a laying factory, since every organ of her body contributes its part to the making of an egg.

The average commercial flock is penned up indoors all year round and never gets the feel of grass or earth. Some of them are in factory-like structures two or three stories high. The chickens do not get much sun or natural light, and are usually saturated with a stale odor of manure and ammonia. The low quality food contained in some purchased mashes doesn't help either. These mashes are made up of formulae designed to high-pressure the chicken into laying her head off. I have seen commercial flocks where all the hens are actually bald-headed. Is this a symptom of excessive egg-laying? I was amused by the advertisement of one of these chicken *entrepreneurs* whose beautiful brochure showed some imposing structures. Under them appeared the caption, "The two triple-deck laying houses at Dash-Dash Farms. Each one is the home of 3,500 breeding birds. These views show not only the houses, but their beautiful location in the scenic Dash-Dash Valley." Do the hens appreciate the view?

Worse yet are the real mass-production factories where the hens are penned up in batteries or tiers of individual cages. Take the case of Mr. X. His *modus operandi* is to set up a number of tiers of chicken coops, in a city—anywhere—just so the location is convenient for shipping to the market. Abandoned factory buildings serve the purpose excellently, and Mr. X once rented a dance hall right in the busiest section of a large city.

Under every tier of coops runs an endless belt, disposing of the chicken droppings which are later sold for so much a bushel. A trough carries water through each coop. As soon as daylight wanes electric lights go on. Experiments show that hens lay 5 per cent more eggs to music—especially organ music, so radios soothe each fowl to do her utmost in egg producing. (It has been found out that they do not like speeches on the radio!)

The newly-laid egg moves down the inclined floor of the coop, into a pocket in front. There is a score pad on each coop, and when the eggs are gathered each fowl who has made good is given a check mark. If a hen lays anywhere from 12 to 16 eggs a month she is in good social standing, but if she drops below 12 she meets the axe.

Broilers also fit into Mr. X's mass-production scheme. Here is where things are reduced to their lowest common denominator. Chicks go in at one end and emerge eight weeks later as broilers ready for the market. An extreme case? No, there are hundreds, if not thousands of them, providing eggs and chickens for urban markets.

A brewery in Detroit, without windows, was a short time ago made into a poultry factory turning out 7000 broilers in a week. Cod-liver oil is used as a sun substitute to prevent rickets. Due to the prevailing system of marketing, such chickens and eggs may land up in the finest grocery stores.

When chickens are "on range," that is, free to run on and eat grass, their eggs have a darker yolk with a greenish tinge. The market considers this as a lowering of the quality and actually penalizes the producer by paying him a lower price! For there are fashions in eggs as in clothes. It is rather ironic that the public by its own fickle whim encourages practices which give it foods of dubious nourishing qualities.

Cannibalism is a vexing problem with chickens—although we do not have it with organically fed flocks. The chickens peck at each other till they draw blood. The sight of blood makes them peck even more. Many deaths occur as a result. The sundry devices that have been designed to prevent cannibalism are clever and even amusing. For instance, there are a dozen different kinds of spectacles. One type prevents the chicken from seeing in front. Another has a red lens so that she cannot distinguish the sight of blood. Whole flocks are fitted with them, believe it or not. I wonder what our grandmothers would have thought of such a scholarly looking flock.

One authority advises the use of much salt in the diet to control cannibalism. Another says "dissolve permanganate of potash in the drinking water until it is the color of cherry juice, and give them no other drinking water." One fellow uses motor oil, lamp black, some kerosene, turpentine and creosote to paint over the pecked part to neutralize the bloody appearance. The "doctors" seem to disagree.

The chemists had a convention at Pittsburgh in the early part of 1944. One of the things discussed was a new chemical process to make tough old roosters taste tender. Pellets of a female hormone called *diethylstilbestrol,* produced synthetically very cheaply, are injected under the skin of the old patriarch, and in from two to six weeks he loses most of his tough roosterishness, and his meat is reported to be tender and succulent. The chemists warned each other, however, that if a piece of meat were eaten in which the pellet had not been fully absorbed by the rooster, the rooster eater would get sick. To protect the public, tough old birds so tenderized should bear a label stating: "This ancient rooster has been fattened and tenderized by the injection of *diethylstilbestrol* pellets under the skin. Eater beware! Before eating be sure that the pellets have completely disintegrated." But they aren't labeled that way.

Mother Nature has a way of caring for her children. The hen must have a rest period, a period of lower production, to build herself up. So hens moult, usually in the early part of the summer, and lose feathers. Laying is at low ebb at that time. But poultrymen are now even interfering with this natural check. They say that this period can be postponed or reduced and they do it by certain feeding practices. A hungry hen is apparently even crazier and sillier than one trying to cross the road in front of your car.

Lighting up the hen-house at night is done by many flockowners in the winter time when darkness comes early. At this season of the year egg prices go up because of reduced supplies. By lighting up the loft, daylight conditions are simulated and the chicken continues to eat. The more she eats the more she lays. Egg authorities say that this practice doesn't increase one whit the total number of eggs laid by a chicken in a year. What it does is to equalize laying over the different seasons and thus increases the income of the business. What it does to the health of the chicken, who should work in harmony with the rhythm of the seasons, is not given a thought.

Another artificial method that is beginning to be heard of is the 26-hour day instead of 24. Since hens lay eggs every 26 hours instead of every 24, it is believed that the proper thing to do is to lengthen their day by two hours. So they are placed under artificial light and permitted to sleep for 12 hours. They are kept awake for 14 hours, irrespective of the time "outside." Other poultry professors believe that egg-laying in hens can be speeded up by injections of hormones. This has been tried, and it has been found that the periods between the laying of eggs can be reduced to 17 hours instead of 26 or 24.

As a result of all these practices the general bodily resistance of the average hen today is very low, and disease among chickens is rampant. Poultry scientists are at their wits' end diagnosing and attempting to keep such diseases in check. At many farms you will see some dead and paralyzed chickens around the laying houses all the time. Instead of attacking basic cause and giving the hen food that is raised properly and which would keep her healthy, these poultry pathologists approach it in the most "scientific" and "professional" and complicated manner.

They advise vaccination against fowl pox, larynotracheitis, and other diseases. This is very elaborate and you must call on the veterinarian to choose the proper vaccine and to describe how and when to vaccinate. There are hundreds of patent medicines on the market, and they wouldn't be there if the poultryman did not make regular use of them. There are vitamin mixtures, craw conditioners, patent laxatives, deworming mixtures, *ad infinitum*. This is hardly the place to describe the hundreds of diseases and pests which plague poultry but the disinterested public would get the shudders if it could open up one of the tremendous tomes devoted to diseases in chickens and see the hideous pictures of some of them in the last stages of incurable disease. You eat eggs that come from some of these sick birds.

The 1942 U. S. Yearbook of Agriculture, page 1007, carries the statement, "losses from poultry diseases in the United States have been estimated by Government authorities to be approximately 100 million dollars annually. . . . A few years ago poultrymen in many sections of the United States were forced to abandon the raising of turkeys because of the prevalence of blackhead, caused by a protozoan parasite. . . . Farmers are losing 18.8 per cent of their poultry because of disease, or 1 out of every 5 birds." When you add to this 100 million dollars, the value of the farmer's labor in combatting disease, the cost of medicines, vaccines and preparations, the amounts paid to veterinaries, and the cost of research at Government Agricultural Colleges, etc., the total might come to at least twice that sum.

Sir Bernard Greenwell in a paper published in the *Journal* of the Farmer's Club in 1939, describing feeding tests which he made in England, said:

> "The grain from fertile soil was found to contain a satisfying power not produced by ordinary produce. But this was not all; resistance to disease markedly increased. In poultry, for example, infantile mortality fell from over 40 per cent to less than 4 per cent. In pigs, troubles like scour disappeared. Mares and cows showed none of the troubles which often occur at birth."

We can see the results of organically grown feed in our own chickens which are beautiful birds, admired by all who see them, and full of energy. We have very little death from disease. A great part of our success is due to the fact that our chickens get a daily ration of earthworms which we raise for that specific purpose.

From a nutritional standpoint, the best eggs come from chickens that can run on grass in the spring and summer and have access to the outside all year round, supplemented by a diet of properly raised food. On our farm we feed only food grown on our own place with the exclusive use of natural organic composts. Purchased mashes are taboo. The ration consists of corn, wheat, sunflower seeds, barley, buckwheat, soybeans, rye, alfalfa, and earthworms bred for that purpose. The chickens thrive upon it and, as I have said, know practically nothing of disease. In addition, in fact most important of all, the meat has a wonderful taste.

My wife brought a cooked chicken to a sick friend who called up a few days later to compliment her on her ability to cook. The friend said she had never tasted such delicious chicken and wished she could cook as well. When she was told it wasn't the cooking but the chicken, and when we proved it to her by giving her an uncooked one, she was pleasantly surprised and convinced.

One of the principal reasons why the public gets poor quality eggs and chickens is the low price the farmer gets for his product. Given a fair profit more poultrymen might take pains to turn out an eatable egg and chicken.

There is a tremendous market waiting to be developed—the production of high quality eggs by chickens kept under natural, healthy conditions, at a price higher than the average. People will gladly pay higher prices if they know that their health will be benefitted and their palates rewarded.

Dr. Oliver, previously mentioned in connection with earthworm breeding, was also a pioneer in improving the chicken-droppings pit.

In cleaning the dropping boards it is a rare thing to find a 100% job done. Some droppings remain. The day after the scraping is done, the boards again become covered. Sometimes, due to pressure of other work, the cleaning is overlooked for weeks. But even where the cleaning is accomplished each week there is practically always an overpowering smell. I recently visited the chicken lofts of a large-scale chicken raiser whose product is considered tops in the community. The sickening odor was almost enough to knock me down. Apart from its nuisance as an offensive odor, the escape of ammonia from chicken manure represents a big loss of nitrogen from a fertilizer standpoint.

For many years the more progressive poultrymen have been using a pit instead of boards to eliminate the work of cleaning each week.

The droppings fall into this pit and do not have to be cleaned out for possibly six months at a time. The odor is kept down by applications of superphosphate. The only trouble with this method is that the superphosphate must be applied almost every day and, as we have seen, it is a strong chemical fertilizer, and endows the chicken manure with all its disadvantages.

By accident, I discovered a method which accomplishes good results. It came about through our way of handling steers. As we do not have fences on our farm, we do not pasture the steers but keep them in the barn, with access to an open, sunny yard. They go in and out as they please and wander all over their portion of the barn. The barn floor is covered with straw. About every two weeks it is our practice to clean out the mixture of straw and steer manure and put it to compost.

In the summer of 1942 the farm labor situation was so acute that we could not find time to clean out the barn and had to revert to the custom of putting down new straw when we normally would have cleaned. As time went on, the steer manure began to rot from the bottom and gradually turned into a good, plentiful compost, which we applied to our corn fields with excellent results. The only drawback resulting from the change was the strong, objectionable odor which arose in the barn.

The following year, 1943, we were forced to follow the same system with the steers but we began to notice something unusual. There wasn't an odor in the barn. Even after four months, all we could notice was the fine, fragrant aroma of straw. *That* was something to think about! The only logical answer seemed to be that as the manure had rotted the bacteria had multiplied, and when we cleaned out the steer-barn we evidently hadn't cleaned it out 100%. This was possible because the concrete floor was very rough and couldn't be scraped clean. In one section, the floor was of earth, and this must have become thoroughly saturated with bacteria, which began immediately to work on the new manure and must have had sufficient power to break it down very quickly. In making a compost heap, there should never be any odors. If there are, then it is a sign that something is wrong and that the heap should be turned immediately. The fact that there were no odors in the barn that second year showed that a regular composting process had taken place.

Well, if the old manure could break down the steer droppings quickly without an odor, why couldn't it do the same thing with chicken droppings? So we made a pit under a second flock of about 40 hens, and in it we put about 6 inches of very fine topsoil. Next we obtained a barrow load of thoroughly rotted steer manure from the very bottom of the manure in the barn and spread it over this soil. The chickens have now been dropping into this pit for several years, and there is not a trace of smell. We clean it out about twice a year. It is necessary to put water in occasionally, but it must not become too soggy. Drainage should be provided at the bottom with at least two or three feet of little stones under the floor level.

I later discovered that this principle had been used in England for centuries—see discussion on Yorkshire foldyards in section on, "Short Cuts in Compost Making" (p. 75).

The advantages of the pit method are first, that it eradicates the odor. I always get a thrill from the reaction of visitors who have smelled other chicken lofts. Secondly, it is by far much healthier for the chickens, sharp odors causing respiratory diseases especially. Thirdly, the chicken manure is composted while it accumulates in the pit. When it is later hauled out it may be used immediately.

Recently we have improved the process by making real compost in the pit. Chicken manure by itself is not a complete fertilizer. Every week or so we place a thin layer of green matter, earth and lime in the pit. I believe the bacteria will break this new matter down without turning. Time will tell. An important aspect of the commercial chicken business is the question of the health of the chickens who must live in the usual putrid atmosphere as well as the health of people who have to eat such chickens and their eggs. I have heard people remark that chickens they have purchased smelled of the barn-yard. Recently we killed 6 of the chickens of the first group that roosted over our compost pit. They were the most delicious chickens we ever ate. The flesh was a delicate texture and marvelously full-flavored. There are two other reasons why this chicken meat was of good quality. First, we had these chickens running on grass during the summer.

Second, we fed them on home-grown feeds raised with organic fertilizers.

Is it possible that one of the important reasons for the increase in disease is the putrid odor which modern fowls bathe in, in present day chicken houses? In a highly odorized chicken loft, one gets the distinct smell of ammonia. This must be especially irritable to a chicken's eyes. In the February 5, 1944, issue of the *Rural New Yorker,* Clarence E. Lee states that ammonia gas coming from chicken droppings is highly irritating and that it has been thought it has frequently led to respiratory troubles in fowls.

Coming back to the dropping pit, its construction may be of wood, brick, cement or hollow tile. Results will pay the cost in a short time. Instead of rotted steer or cow manure as an activator for the pit, where this is not available, use finished compost or scrapings from a place where you have made compost for several years. When you clean out your chicken droppings pit, be sure to leave some at the bottom as a permanent culture.

Earthworm Fed Chickens

A very interesting case has come to my attention from Mr. A. F. A. Konig, of Newtown, Conn., who writes:

> "Regarding earthworms and chickens the following might be of interest to you. Late in 1941 we did some grading in an old buck (goat) pasture which yielded a lot of worms. It so happened that we had several birds which for some reason had lost the proper use of their legs. We disposed of all but three which we put on a worm diet. Two recovered after about two weeks if I remember correctly; the other bird was either accidentally killed or butchered, I forget which."

Some time after that, one of our little chicks about three weeks old became paralyzed. It sat in a corner and would not move or eat. As a rule, a poultryman will not bother with such a case and it will die. Recalling what Mr. Konig had done with three chickens, we went to work and did the same thing. We isolated the chick in a corrugated box and tried to feed it the regular poultry mixture. It positively refused to touch a grain. We then placed an earthworm near its mouth.

You should have seen that little chick gobble it up. We fed it twenty-five earthworms a day. At the end of the fourth day it was running around as chipper as ever. While this instance does not constitute a proof we believe that experiments will bear out the theory.

In the summer of 1943 we conducted an interesting experiment. We took a group of ten Rhode Island Red pullets about 3 months old and divided them into two groups of 5 each. Group A were weighed at the start and were found to aggregate a total weight of 10½ lbs. They were fed the usual kind of diet fed chickens on our farm. Group B which were found to weigh 11 lbs. at the beginning was fed less of the regular ration but were given a daily ration of earthworms.

The experiment lasted about six weeks and the results were interesting. Group B, the earthworm-fed group, put on much more weight and their feathers took on a much richer, deeper hue. You could see that they were more active, lively and healthier looking than group A. Here are the figures on weight.

	Weight at Beginning	Weight at End	% of Increase
A	10½	14½	38%
B	11	17	54%

The 1942 U. S. Yearbook of Agriculture, on page 1033, corroborates our idea of how to build up the bodily resistance of the chicken. It says:

"Animal proteins in the form of milk and meat have been shown to be important dietary supplements in the development of resistance of chickens to *Ascaridia Galli* (round-worms), and a diet wholly of plant origin was not found to be conducive to resistance to helminth (worm) invasion. Ackert and his coworkers have shown that foods high in vitamins A and B increase the fowl's resistance to this nematode and that the lack of the vitamin B complex definitely favors parasitism.'

Earthworms are one of the finest forms of protein for a chicken and the reader can well imagine whether an earthworm contains the necessary vitamins. Nature, in fact, seems to have made the earthworm a specific for the chicken.

There may be a curious explanation for the ready association of earthworms with those other kind of "worms" by persons who have had sad encounters with the latter. Beside roundworms there are gizzard worms, stomach worms, crop worms, tapeworms, trematodes, eye worms, gapeworms, and many others which distressingly afflict poultry. Naturally when the word earthworm is mentioned to anyone who has had a tragic experience with parasitic worms there is an evil connotation with psychological implications in the mere mention of a similar sounding word. Otherwise, they have nothing in common.

From our experience, and that of countless generations of farmers before the chemical era, chickens fed on good food, organically grown, and allowed to range over a wide area for earthworms and insects, are healthy, can and do resist the many diseases that afflict the confined chickens of the commercial operators, and lay eggs that are notable for their food value and excellent taste.

Much study needs to be done—research and controlled experiment—to prove why this is so. I have merely given a few clues from my own experience, and suggestions for raising chickens with far less trouble and nuisance.

Every farm, particularly an organic farm, should have chickens. And the larger the farm, the more chickens can be raised. Let us, in this, as other respects, value quality as much, at least, as we do quantity.

Conclusion

Conclusions should look to the future. And the future of agriculture can be very promising. In almost all quarters of the globe, the kind of commercial farming practiced very generally for the past fifty years is being challenged by practical farmers, soil biologists and other scientists of the agricultural schools, by doctors who have made a study of nutrition, by specialists at work in tracing the origins of animal and plant diseases, and by conservationists interested in keeping our natural resources available for generations to come.

So long as our own American land seemed endless, with an ever-expanding frontier to the West, with new land for the taking, practically, the problems arising from soil depletion through bad farming practice seemed unimportant. People felt they could afford—with a continent to develop—to wear out a farm and move to another. That day has passed. With worn-out acreage, abandoned farms by the many thousands, dust-bowls of frightening area, "tobacco roads," and food that has to be fortified with vitamin pills to fend off illness, we are aware, as never before, of the deficiencies of "modern" agriculture, and are thinking more seriously of the complex nature of soil, the physical requirements of plants and animals, and the re-designing of our fundamental conditions of life than we have been wont to do in the past.

Badly eroded, worn-out soil will not recover over night, but fertility *can* be restored. All of our waste places can be garden and farm and wood-lot again, with time and intelligent handling. Farms that are on the way down, not completely worn out, can be more quickly improved; it is often surprising how quickly land can come back into "good heart"—a secret that only the compost farmers know. Land still fertile can be kept so, with composts, and be constantly improved.

We are, as I have said, more acutely aware of the deficiencies of our agriculture than ever before, and, happily, there is much to indicate that the trend is toward compost farming and gardening. Most of the leading writers on agricultural problems now emphasize the importance of humus in the soil. The garden editors of the large metropolitan newspapers constantly mention the importance of adding humus material, through mulches, manure, or compost—and one comes across many fine articles on the subject in the largest gardening magazines, farming guides, and gardening encyclopedias, publications of the Department of Agriculture, and those of many schools of agriculture.

Hundreds of thousands of agriculturists, amateur and professional, are practicing compost gardening to a small or large extent, experimenting on their own, in many cases, to see whether it is an improvement over the "scientific" way. The fact that enough American farmers are turning to it to warrant farm-machinery manufacturers in making special machinery for handling large-scale compost installations testifies to the results these farmers are getting and to the success of their own experimenting.

One of the greatest reasons for turning to composts (I am thinking especially now of amateurs and the "marginal land" type of farm, where the farms and villages are poor and without much cash) is the low cost of starting to remodel the land. Just the waste material on the place—the autumn leaves, weeds, marsh hay, table scraps, stable bedding and such animal manure as can be obtained, will make compost. A garden, or one field at a time, or more, depending on quantity, can be brought back to good fertility. No heavy outlays for expensive and strange materials. Mulches will help, if you want to start before your compost is ready.

The possibility of starting "from scratch," at little or no cost, is a most important feature, for it depends on nothing more than the resources of the farm or home place itself; we do not need to wait for a government program, national resources surveys, subsidies—or tack on another mortgage—to begin.

This is borne out in a most interesting article in the June, 1945 *Survey Graphic* (condensed in the July *Reader's Digest*) by J. P. McEvoy, describing the work D. Spencer Hatch is doing in a small valley about fifty miles from Mexico City. He has set up a one-man model farm to show the Indians of that area how to improve their land and homes. Mr. Hatch doesn't teach or preach, but just "does." To quote from McEvoy's article:

"It has been said that the only crop that can be raised on poor land is poor people. Poverty, ignorance, and disease, in Mexico as elsewhere, are inseparable problems that must be tackled all at once, and from the bottom. Hatch started at the bottom—with the soil, worn out by centuries of misuse.

"He built up a piece of wretched ground with compost made of plant life and natural fertilizers available to the poorest farmer, and grew corn twice as high and four times as productive as his neighbors'. In Mexico, where pests ravage every one of the Indians' nine varieties of beans, Hatch's beans flourished with nary a bug to bite 'em. On his other demonstration plots, small fruits and herbs unfolded month after month in a succession of miracles.

"In three years, Hatch transformed a worn-out valley into a tiny paradise of rejuvenated soil and bumper crops."

His example is widely copied by the Indians. As Mr. McEvoy points out:

"Here, at amazingly low cost, Hatch has developed crops, fruits, livestock, and living amenities which eventually should benefit most of Mexico. Each building, from the smallest chicken house to the home in which Hatch and his family live, demonstrates how to use effectively the cheapest materials within reach of the lowliest peon." So impressive is Hatch's work that, according to the article, "The Rt. Rev. L. G. Ligutti, Secretary of the National Catholic Rural Life Conference, came twice—the second time with the President of the Conference, Bishop Schlarman of Peoria, who said, 'This is a *must* for all priests who go out to do rural work in this part of the world.' "

The fact that conversion to compost farming can be done simply, starting with a mulch or by sheet-composting (which is only a mulch plus manure), and at little or no expense makes it the simplest and most universal of programs. No vast amount of chemical or bioiogical learning is necessary; it is a program that holds good with any soil, be it alkaline or acid. I am not in favor of ignorance—but could the Mexican Indians who are learning from Hatch's model farm manage to learn the intricacies of chemicalized agriculture? Could our own Okies, or Tobacco Road "poor whites"? Can anyone, without a long serious study of chemistry, for which most amateur gardeners and farmers are ill adapted?

If I have stressed the value of compost gardening to amateurs occasionally, in this book, it is certainly not because I think the professional farmer can overlook organic agriculture; quite the contrary—this book is addressed chiefly to the general farmer. But it seems to me that millions of Americans who would normally be tending a small garden spot, or would have a "week-end farm" have been scared off the land by the intricate technology of chemicalized agriculture—and by the problems that this book has dealt with. Farming has come to seem, in recent years, too complex, too unrewarding. Erosion, soil exhaustion, fighting the insect pests with a dozen sprays, stiff outlays for expensive fertilizers, and poor crop results, have made the game not worth the candle. After a season or two, the amateur gives up. Professional and business people, who commute to our large cities, all too often, give up their "part-time farming" attempts for the same reasons, settling at last for a few ornamental shrubs and flowers, and a staple or two like potatoes and tomatoes, letting whole farms go back to "pasture"—more literally, scrub brush—when it is not at all necessary or desirable. Many of these people, discouraged by the complexity of "scientific agriculture," could have enjoyed rural life and good husbandry, and could have found, like Washington, that farming was their "chiefest amusement."

The country, of course, depends on its professional farmers for its main food supply. But in the years directly ahead, every agricultural resource will be strained to the utmost to feed the battered world, and all the small gardener and amateur can produce will help the food supply, at least locally.

I have mentioned amateurs so much in this chapter and shall continue to do so because I believe in farming as a way of life, and that the joys of it have been almost completely lost in practice through the great difficulties, frustrations, expense, and havoc created by chemicalized agriculture.

With veterans of the war returning, and wanting, many thousands of them, to become farmers, it is all important, it seems to me, to let them know that there is a way of farming that is, on the whole, pleasant and rewarding, one that does not require a long course of chemical study by way of preparation. In farming with composts, the new farmer can in a relatively short time get his land into decent shape, fertile and productive, with many of the pitfalls of "modern" agriculture largely or completely by-passed. He will find it an agreeable career and way of life. If returning veterans who are new to farming go back to the land and try to farm the "scientific" way without technical chemical education, they are almost sure to fail, after a great deal of hard work and heart-break.

The general pattern of living, I believe, is bound to change in America, very quickly and increasingly in the next few years. During the war industry has been greatly decentralized. Factories have been built in rural areas and in small towns "near the land." Population has shifted enormously—and there are hundreds of thousands of people who will not go back to the congestion of the swollen cities. The freight-rate equalization recently won by the South and West will mean a tremendous expansion of the small-town and rural areas of most of the United States. Communities are planning along new lines, and into these plans the program of compost gardening fits beautifully, to aid in better living.

Paul Corey in *Buy An Acre*—a good, practical handbook for the amateur who wants to have a small country place—predicts that, after the war, "The country within a radius of from fifty to one hundred miles of our cities will become the New Frontier of America. Ten million tiny homesteads, each with an acre or so of ground on which to raise a few chickens and the family's yearly supply of fruits and vegetables, will spring up within commuting distance of factory and business. Congested urban and industrial areas will eventually dissolve over the land. . . .

"This time millions of people will go to the land for homes—an acre, two acres, never more than ten. They will bring this dead land back to life again, build strength again into the soil. The first time we took the land to exploit it; now we are taking the land to save it." *

* Dial Press, 1944. Condensed in *Reader's Digest*, July, 1945.

This may seem an overstatement, but the movement has actually begun on a large scale. The overcrowded cities have driven people to look far afield for housing—and the urban exiles are discovering something in the way of new worlds, and new ways of enjoying life.

The back-to-the-land movement is a formidable one—whether you construe it to mean to farms for careers in agriculture, or whether you imply a sort of small town garden developments and "week-end" farms for commuters. It is a fact that the majority of human beings need some contact with country life, to restore the soul.

Regional planners are taking this into account, trying to develop the new industrial sections so that the community life will be a good blend of town and country. It is as if William Penn's hope for the Philadelphia he laid out, each house with its garden and orchard, as "a good green towne," were to serve as a model at long last.

In the July, 1945 issue of *Theatre Arts,* Clarence S. Stein talks of the future theatre in terms of the same "dissolving city" era that Paul Corey predicts. Let us look at the picture he draws:

> ". . . I believe that the cities we know will practically disappear to be replaced by the Regional Cities of the future. From the sky, the Regional City will not seem a great city at all. No endless man-made mass of structures. Only a constellation of small or moderate-sized towns, separated by great areas of natural forests and fields, farms and recreation grounds. The Regional City will be a green city—not a gray one. Even the towns will be park-like. Buildings will be designed in close relation to their natural setting. They will have a unity and beauty that comes from an integrated setting developed for effective and joyful living.
>
> "Though separated by great open spaces, permanently preserved for farming, recreation or forest, these communities will be bound together by express park highways. Thus, in time-distance, most people will be closer to professional theatres, university playhouses and other places of entertainment and education, than they are today."

One hopes, truly, that these visions of the future may come true. They seem realistic enough, for many straws seem to be blowing in the same wind.

This back-to-the-land movement is one of the healthiest signs of a progressive people. Civilizations that get too far from the land are bound to decay. Nations, like Antaeus of old, need to renew their strength by contact with the earth. Town and farm complement each other, need each other, and must understand each other's problems and ideals. If not, society and civilization decay through ignorance, antagonism, exploitation and the artificiality that is their product. The back-to-the-land idea is a healthy sign that our people know their need for country life. This New Frontier idea is sound, for it advocates that as many people as possible live on the land—a means of preserving a sturdy, healthy population, under sound farming practices.

The work of other individuals and groups is notably changing, in slow and healthy ways, the rural environment of America, as part of a realistic back-to-the-land idea, honest and factual, divorced from fantasy and wishful thinking. Ed Robinson's Have-More Plan is along the lines of Corey's New Frontier, and gives good, sensible advice to the commuter who wants to cultivate a small-acreage home-place for exercise, hobby and better living. Likewise, The School for Living at Suffern, New York, has developed many valuable techniques for making rural life and home industry and arts exciting and profitable.

Federal and State government agencies, 4-H Clubs, Farm Bureau, Grange, many rural cooperatives, and county libraries are carrying out excellent programs in many communities throughout the nation—programs of education in agriculture, home economics and handicrafts, and in recreation that are doing a great deal to aid in solving the problems the farm family faces, and in making rural life richer in social opportunity.

The National Catholic Rural Life Conference, mentioned before, is perhaps less concerned with material objectives of an economic sort than with the stuff of human living itself, but it is one of the great forces helping to create a vital back-to-the-land movement.

To quote from the article, "Why Rural Life?" by Edward Skillin, Jr., in *The Commonweal.*

"What is the ideal the National Catholic Rural Life Conference holds forth? On the one hand there is the family-centered farm—ordinarily not more than 40 acres and often less. Here a man and his wife and children devote themselves primarily to becoming self-sustaining as to food and feed, with certain crops and handicrafts produced for the sake of meeting cash needs. Such a farm intelligently selected and managed provides a security for the family far beyond that possible in the uncertain industrial world. Children are an economic asset, since food is plentiful and they can begin to help around the place at an early age. Furthermore no environment is better than the farm for developing initiative, responsibility, manual dexterity, creativeness, appreciation of the beauties of nature, and, if it is a component part of a good rural community, good social attitudes.

"In the second place the Conference is strong for part-time farming. As First-Vice-President Rev. Patrick Quinlan of Brookfield Center, Connecticut, explained it at one of the sectional meetings [at the last annual meeting, at Cincinnati], a man should have at least an acre of arable land to qualify as a part-time farmer. This will enable him to reduce his living costs by producing all the vegetables he would otherwise have to buy at the store. It may even enable him to produce a small cash crop. All this is particularly useful if he works in a seasonal industry and it serves as a real cushion whenever he is laid off. He has room and light and air to raise a family, while the cultivation of a plot of ground of this size enables him and his family to develop the rural virtues outlined above. If he does it well, he may eventually want to devote all his time to farming, and almost certainly his children will develop that love of the land which may lead to their going in for the real thing—when they reach adulthood—full-fledged rural life on the small farm."

The Conference publishes a most interesting monthly magazine, *Land and Home,* full of suggestions for making rural life more rewarding socially and spiritually, as well as practical articles on good, sound farming.

Another group of pioneer thinkers in regenerating American rural life, Friends of the Land, has grown from sixty members in 1940 to 7,500—a group of progressive men and women among bankers, writers, soil scientists, conservationists, gardeners, artists, county agents, teachers, farmers and others greatly interested in revitalizing country communities. Their quarterly magazine, *The Land,* always is full of valuable, stimulating articles, placing it up among the best.

In quoting from Liberty Hyde Bailey's article, *Marvels at Our Feet,* in the Spring, 1945, issue of *The Land,* I believe I am not incorrectly giving the ideal not only of the Friends of The Land, and the Catholic Rural Life Conference, but of all the honest back-to-the-land pioneers:

"We may as well, I think, accept it first as last that the farm is not the place from which to derive great money incomes. We should soon be ready to turn from our false gods and to capitalize the riches that may belong to a farmer, or to a gardener, and to nobody else. The art by which from day to day one may grow, oneself, in country living; the appreciation of a separate life; the love of beautiful surroundings and of all those gracious solitudes that lift the inner mood—without these there is really very little incentive to live in the country; without these country life is tame.

"It is not enough to be comfortable and make money. There is no satisfying project that does not run beyond the cash in hand. There is no relish in households not well managed, no joy in gardening in gardens not well kept, no gratification in work not well performed. And there is no pleasure in leisure that is barren of wonder and unproductive of those enthusiasms that warm the heart and burn in the brain."

The opposite trend, which should be mentioned again, and perhaps called the "away-from-the-land" movement, is the "chemurgy" idea, extending large-scale monocultural practices to supply industry with agricultural materials, and the many-thousand-acre farms for single-crop items like wheat and cotton. This collectivized plantation kind of commercial farming ends in disaster to the land, as we have pointed out, if safeguards are not set up and maintained, to insure a living soil. Composting, crop-rotation, fallowing and green-manure crops are essential, if this way of supplying industry is to be anything but a long-range disaster. Our destruction of forests, creation of dust-bowls, ruining areas larger than many entire countries through land-exhaustion and erosion, must stop. This assembly-line, machine-run agriculture, depopulating rural areas or making them insufferably lonely, is very much a trend in some parts of the country. Do we want its "deserted villages"? These evils are not inevitable, or natural things. If we, as a nation, permit these practices to go on, we shall richly deserve consequences such as those predicted by the prophet Micah: ". . . the land shall be desolate because of them that dwell therein, for the fruit of their doings."

But I believe we shall not be stupid enough, as a nation, to let the vicious land-mining practices continue much longer. We are now aware of their danger, and Americans are noted for their good common sense. We have come, in recent years, to see the terrible waste and tragedy of exploiting the land to our ruin; every intelligent and progressive force within the nation is on the alert, knowing that exhaustion of the soil leads to exhaustion of a people. Probably the most severe criticism that can be made of "modern" commercial farming is that it has, for various reasons and in many ways, driven people from the land, or turned so many into a "rural proletariat"—the hopeless share-cropper, hill-billy, shiftless "renter," or migratory worker.

The coming years will, if I may make a prediction, be the most exciting ones known in agriculture and country living for many decades. The pioneering work of Sir Albert Howard and other soil biologists is just beginning to bear fruit in practical farming, and is serving as an inspiration to research in many facets of agriculture and medicine. The science of nutrition is really in its infancy, with new discoveries—such as those of Dr. McCarrison—coming to light. Medicine has made dramatic and wonderful use of penicillin and opens up great new vistas in use of soil bacteria.

I believe a whole new era of agricultural research is in the making—one that will benefit the country at large far more than all the research of the past has done, one that will more nearly help to create a healthy society and keep it in close touch with the land from which it gets its strength and sweetness—a country without city or rural slums, a country of homes and gardens, parks and forests, a country of prosperous farms and a healthy, vigorous people creating a fine, new community life "in the pursuit of happiness."

BIBLIOGRAPHY

MAGAZINES

The Compost Club Magazine, Box 1303, Auckland, New Zealand, issued 6 times yearly. 5 shillings per annum.

Land and Home, issued quarterly. Des Moines, Iowa.

Soil and Health, issued 3 times a year. Edited by Sir Albert Howard, 14 Liskeard Gardens, Blackheath, London, S.E. 3. Subscription 5 shillings per annum. Single copies 2 shillings.

Organic Gardening, Emmaus, Penn., issued monthly. Subscription $3.00 per annum.

The Farmers Weekly, 42 East Burger St., Bloemfontein, South Africa. 25 shillings per year.

The Land, issued quarterly. Bel Air, Maryland. $5.00 per year.

BOOKS

Soil and The Microbe, Waksman and Starkey. John Wiley. New York, 1931.

Plowman's Folly, Edward H. Faulkner. Univ. of Okla. Press, Norman, Okla.

Principles of Soil Microbiology, Selman A. Waksman. Williams & Wilkins, Baltimore, 1932.

Soil Conditions and Plant Growth, Sir. E. John Russell. Longmans, Green, New York, 1942.

The Nature and Property of Soils, Lyon and Buckman. Macmillan, New York, 1943.

The Living Soil, Lady Eve Balfour. Faber & Faber, London, 1943.

The Soil and Health, Sir Albert Howard. The Devin-Adair Co., New York, 1947.

An Agricultural Testament, Sir Albert Howard. Oxford Univ. Press, London, 1940.

New Garden Encyclopedia, (Ed.) E. L. D. Seymour. Wm. H. Wise & Co., New York, 1942.

A Popular Treatise on the Physiology of Plants, Paul Sorauer. Longmans, Green, London, 1895.

Bio-Dynamic Farming and Gardening, Ehrenfried Pfeiffer. Anthroposophic Press, New York, 1940.

Compost, F. H. Billington. Faber & Faber, London, 1943.

The Garden Dictionary, (Ed.) Norman Taylor. Halcyon House, New York, 1942.

Trampling Out the Vintage, Joseph A. Cocannover, University of Oklahoma Press, Norman, Okla., 1945.

Pleasant Valley, Louis Bromfield, Harper and Brothers, New York, 1945.

Buy an Acre, Paul Corey, The Dial Press, New York, 1944.

Forest Soils, S. A. Wilde, Chronica Botanica Co., Waltham, Mass., 1945.

Plants and Vitamins, W. H. Shopfer. Chronica Botanica Co., Waltham, Mass., 1943.

Farmers of Forty Centuries, F. H. King, Jonathan Cape, London, 1933.

The Cancer Problem, Dr. C. E. Greene. Greene & Son, Ltd., London, 1917.

The Compost Gardner, F. C. King. Titus Wilson & Son, Ltd., Kendall, England, 1943.

Legal Medicine and Toxicology, R. L. Emerson, M.D. D. Appleton & Company, New York, 1909.

Commercial Fertilizers, Gilbert H. Collings. Blakiston Co., Philadelphia, 1941.

Soils and Fertilizers, Bear. John Wiley & Son, New York, 1942.

The Story of Man and His Food, C. C. and S. N. Furnas. New Home Library, New York, 1937.

Weed Control, Robbins, Crafts and Raynor. McGraw Hill, New York, 1942.

Eat, Drink and be Wary, F. J. Schlink. Covici, Friede, New York, 1935.

The Practical Flower Garden, H. R. Ely. Macmillan, New York, 1911.

The Soils That Support Us, Charles E. Kellogg. Macmillan, New York, 1941.

The Clifton Park System of Farming, Robert H. Elliott. Faber & Faber, London, 1943.

Humus, Waksman. Williams & Wilkins, Baltimore, 1938.

History of Civilization, Henry Thomas Buckle. J. W. Parker, London, 1857.

Cancer and Diet, Frederick L. Hoffman. Williams & Wilkins, Baltimore, 1937.

History of Randleigh Farm, Ernest Scott and L. A. Eif.

Science in the Garden, Harry Britton Logan. Duell, Sloan & Pearce, New York, 1941.

Insect Invaders, Anthony Standen. Houghton, Mifflin, Boston, 1943.

100,000,000 Guinea Pigs, Kallet and Schlink. Vanguard Press, New York, 1933.

Irrigation and Drainage, F. H. King. Macmillan, London, 1899.

Microbiology, Charles Edward Marshall. Blakiston.

Tomorrow's Food, James Rorty and N. Philip Norman, M.D. Prentice Hall, 1947.

Smith & Wesson Hand Guns.....McHenry & Roper
Hunting Pronghorn Antelope.....Bert Popowski
The Woodchuck Hunter.....Paul Estey
Small Bore Rifle Shooting.....Edward Crossman
Modern Bow Hunting.....Hiram Grogan
Salt Water Fishing Is Easy.....Jerry Sylvester
Fishing Flies and Fly Tying.....William Blades
This Business of Exploring.....Roy Chapman Andrews
Handbook for Shooters and Reloaders.....P. O. Ackley
Greatest Fishing.....Joe Brooks
Hunting Our Biggest Game.....Clyde Ormond
How To Hunt Deer.....Edward Freeman
Burning Powder.....Douglas B. Wesson
Practical Rifle Shooting.....Walter Winans
Handloading Ammunition.....J. R. Mattern
Shooting.....J. Henry Fitzgerald
Rifles & Rifle Shooting.....Charles Askins
Mauser Rifles and Pistols.....W. H. B. Smith
The Book of the Pistol & Revolver.....Hugh B. C. Pollard
Woodcraft for Women.....Kathrene G. Pinkerton
Woodcraft.....Owen Jones & Marcus Woodward
The complete Journals of Henry David Thoreau

Made in the USA
Monee, IL
14 September 2026

62974516R00152